Seabirds & Other MarineVertebrates

Competition, Predation, & Other Interactions

Seabirds
& OTHER
Marine Vertebrates

Competition, Predation
and Other Interactions

EDITED BY
Joanna Burger

Columbia University Press   NEW YORK

Columbia University Press
New York   Guildford, Surrey
Copyright © 1988 Columbia University Press
All rights reserved

LIBRARY OF CONGRESS
Library of Congress Cataloging-in-Publication Data

Seabirds & other marine vertebrates : competition, predation & other
  interactions / edited by Joanna Burger.
  p. cm.
    Includes bibliographies and index.
    ISBN 0-231-06362-8
    1. Sea birds—Ecology. 2. Marine fauna—Ecology. 3. Marine ecology.
I. Burger, Joanna. II. Title: Seabirds and other marine vertebrates.
  QL673.S28 1988
  598.4—dc19   88-11698
                                                                    CIP

Book design by Charles Hames

Printed in the United States of America

Casebound editions of Columbia University Press books are Smyth-sewn
and printed on permanent and durable acid-free paper

*For Everyone Who Ever
Lived at 1221
Ferry Road*

■ Contents

■ Preface

Seabirds are unique among marine animals because they spend the majority of their time above the water-air interface. Most seabird species nest on oceanic or coastal islands, although a few breed on mainland areas. Seabird interactions with other marine vertebrates are limited to sporadic, although important, interactions when birds enter the water to feed or when they breed on islands used by other marine vertebrates. The overall aim of this volume is to provide an introduction and representative selection of current research dealing with interactions between marine birds and other vertebrates in marine habitats.

Many marine biologists have concentrated on the species-by-species approach to the study of marine vertebrates. Although it has long been recognized that species interact, it has been far easier to study that interaction from the viewpoint of only one of the participants. Recently biologists have tried to study interactions from the viewpoint of all participants, including examining the adaptive advantages and evolutionary disadvantages of the behavior of both species. This volume will examine what is known of interactions of marine birds with other marine vertebrates. Because of the high visibility of birds, it is easy to observe their interactions with other vertebrates such as fish, porpoises, and whales, but it is difficult to obtain quantitative data on interactions. For many years seabird biologists have described flocks of seabirds foraging over fish schools or groups of mammals. However, many of these observations are anecdotal, or descriptive at best. In recent years seabird biologists have started to relate the behavior of seabirds to the conditions of the ocean waters below, including the presence and behavior of other marine vertebrates. This volume will fill a particular niche: that of pulling together in one place the exciting current work in marine species interactions. It is my feeling that many scientists have just begun to develop paradigms for the study of such interactions.

The majority of the papers herein are structured to include a balance between literature review, original data, and synthesis. The research approaches involve examining seabirds as they interact with other marine vertebrates, but many authors have concentrated equally on the "other vertebrates" as well. The marine habitat provides unique constraints to the study of the interactions of organisms because of its vast size and openness, and the difficulties of working within a water environment with limited visibility or detectability from above or within.

Most of the essays concern obvious aspects of predation, competition, and commensalism. These topics relate directly to food acquisition and survival in a marine interface. They provide paradigms for the future study of marine birds and other organisms. Other important aspects of marine bird behavior, such as predation on breeding seabirds in mainland areas, have not been included because they involve nonmarine animals. Interactions with humans have been included only when they occur in the oceanic waters, but not when they occur only on breeding colonies (such as human disturbances and egging).

Since man's encroachment in coastal and pelagic environments is increasing and poses a threat to marine birds, a volume elucidating various aspects of their interactions has multiple uses. As well as being of value to ornithologists, the volume should be useful to other vertebrate biologists, behaviorists, ecologists, and managers involved with coastal planning.

I am especially grateful to R. G. B. Brown, M. Gochfeld, J. Jehl, C. Leck, several of the other contributors, and several anonymous reviewers for so ably helping to review the manuscript. I thank M. Gochfeld and C. Safina for help in indexing.

■ Introduction

1

■ Interactions of Marine Birds with Other Marine Vertebrates in Marine Environments

Joanna Burger • *Department of Biological Sciences Rutgers University*

Marine birds are unique among birds in that they are capable of existing for extended periods in marine environments, feeding, traveling, and resting away from land. Further, although most fly above the oceans while searching for concentrations of food, they capture their food on or beneath the surface. Thus seabirds coexist with a variety of other, often very differently adapted species, and these interactions vary with the time of day, the season, the weather, their reproductive stage and behavior, and the presence and behavior of other marine organisms. Consequently, their interactions with other vertebrate species, or their mechanisms for coexistence, can be expected to vary dramatically depending on their habitat use and activities.

In the following sections I discuss 1) species coexistence and the mechanisms for such coexistence (competition, predation, mutualism, or commensalism), 2) marine interfaces (open versus closed systems), and 3) types of interactions of marine birds with other vertebrates.

■

Species Coexistence

Wherever species coexist, ecological explanations for their existence include such primary mechanisms as competition, predation, mutualism, and commensalism. Within a trophic level, competition may allow some species to exist while excluding others from an assemblage; and predation impinging on the assemblage may allow coexistence (see examples in Connell 1961; Paine 1966; Harper 1977; Neill 1972). Marine systems are unusual in that opportunities for predation involving vertebrates as both predators and prey can result in very dense aggregations of predators which are simultaneously competing directly in certain ways (e.g., for food items) and facilitating predator coexistence in other ways (e.g., increasing prey vulnerability to predators). Competition occurs when a number of animals (of the same or a different species) utilize a common resource in limited supply, or when animals using the same resource harm one another in the process (Birch 1957). Further, the intensity of interspecific competition is inversely related to the amount of interspecific resource partitioning (Pacala and Roughgarden 1982). Predation can be defined as any interaction in which energy flows from one organism to another (Sih et al. 1985), while commensalism occurs when both organisms benefit from the interaction.

Competition

Initially competition received the most attention as the factor affecting coexistence in communities. Competition theory was developed by Lotka (1932), Volterra (1926), and Gause (1934), followed more recently by MacArthur (1958), Hutchinson (1959), and MacArthur and Levins (1967). Evidence comes from descriptions of interactions in nature and from field experiments (Birch 1957, Schoener 1974). Field experimentation largely dates from the early 1970s (Harger 1970a, 1970b; Wilbur 1972; Connell 1974, 1975; Harper 1977), and the number of studies has increased dramatically since then (Castilla 1981). After a number of papers supported the existence and importance of competition to species interactions (May et al. 1979), the importance of interspecific competition was questioned (Strong, Szyska, and Simberloff 1979; Connell 1980; Schoener 1982), defended (Hairston 1980; Roughgarden 1983), or viewed as a temporally sporadic, often impotent interaction (Wiens 1977). Partially these arguments arose because of a lack of experimental evidence from the field, or because supportive observations were statistically indistinguishable from random events.

In 1983 Schoener reviewed field experiments on interspecific competition. He defined an interspecific competition experiment to be a manipulation of the abundances of one or more hypothetically competing species, including removals, introductions or both, and a field experiment as one in which some major natural factor extrinsic to the organisms of interest is uncontrolled (Schoener 1983). He examined 164 studies, and 90 percent demonstrated some competition. Marine, freshwater, and terrestrial systems showed similar levels of competition. Of the 390 species examined, 76 percent sometimes showed competition, and 57 percent showed competition under all circumstances. For marine species, 56 percent always showed competition. Biases occurred in that most studies were in temperate regions, most were continental, and most investigators chose systems where they expected to find competition (Schoener 1983). Further, I note that the marine studies were largely estuarine or coastal, rather than pelagic.

Likewise Connell (1983) examined interspectific competition over a nine-year period from five ecology journals and the *American Naturalist*. His criteria for inclusion were adequate controls and presentation of data to judge 1) how similar the experimental setup was to natural conditions, and 2) whether or not interspecific competition was occurring. Competition occurred if following an experimental change in abundance of a competitor there was a statistically significant response in the opposite direction in the species being studied (Connell 1983). He thus examined seventy-two papers including 215 species. He found that when only one species was being studied the incidence of competition was 93 percent, but when two or more were studied it dropped to 48 percent, suggesting that the former figure, and that reported by Schoener (1983) above, may be artifacts of investigator selection of study organism. This further reflects the difficulties of publishing negative results, particularly in the more prestigious ecology journals. Generally, interspecific competition was not distinguished from intraspecific competition, but where it was, interspecific competition was the stronger form in one-sixth of all experiments (Connell 1983). Competition was more prevalent in large-sized, compared to smaller, organisms. Competition occurred in 55 percent of the 200 species examined in detail (Connell 1983). This percentage decreased when two, rather than one species, were examined, indicating asymmetrical competition, although reversals in rank order of competitive superiority have been demonstrated (Connell 1983). Interspecific competition for marine studies ranged from 59 percent for plants and 78 percent for herbivores to 75 percent for carnivores. These values were higher than for terrestrial and freshwater studies. But again, marine studies were generally limited to coastal rather than pelagic habitats.

Schoener (1983) expanded the usual concepts of exploitative and interference competition to include:

Consumptive: Use of a resource, thereby depriving other individuals of it.
Preemptive: When space is passively occupied, thereby causing others not to occupy it.
Overgrowth: When an individual grows over another individual, thereby depriving it of light or access to waterborne food.
Chemical: When an individual produces a toxic chemical that harms other individuals.
Territorial: When an individual aggressively defends a unit of space, preventing others from occupying it.
Encounter: When mobile nonspatially attached individuals harm other individuals (through time or energy losses, theft of food, injury).

These mechanisms of competition are useful for species continuously occupying the same niche. However, marine birds, traveling widely across expanses of ocean, may be only sporadic members of particular assemblages, and thus their effective competition may be intermittent. Seabirds may change the foraging behavior of competitors for only brief periods of time or only during brief periods of the year. Some seabirds breed in the interior regions of continents, far from the coast, and do not impact marine environments during their breeding season.

Hairston, Smith, and Slobodkin (1960) proposed that herbivores at a trophic level are not likely to compete for common resources in terrestrial habitats, but are controlled by their predators; whereas at other trophic levels the organisms compete. Schoener (1983) found positive support for this hypothesis (although his sample size was small), but Connell (1983) did not. In marine habitats Schoener's data did not support the hypothesis. A second hypothesis concerning trends in competition across trophic levels is that predators will compete more than herbivores, and herbivores will compete more than plants (Menge and Sutherland 1976), but Connell's (1983) data did not support these hypotheses. Connell (1983) also found evidence that small-bodied individuals are more vulnerable to competition than large-bodied ones. However, this relationship has not been examined in many marine vertebrates.

All of these studies clearly indicated that terrestrial organisms show a lower frequency of interspecific competition than marine ones. However, the marine species reviewed (Schoener 1983; Connell 1983) were mostly intertidal invertebrates. The paucity of marine vertebrate data is largely a result of the inability of marine vertebrates to be transplanted or enclosed without ill effects, a necessary condition for a "field experiment" as defined by Schoener (1983) and Connell (1983). Similarly, few field experiments examined freshwater birds, although Ericksson (1979) showed that competition between carnivorous ducks and fish in Swedish lakes varied annually. Similarly, Eadie and Keast (1982) found that ducks and fish compete for food. However, without marked, confined individuals it is difficult to interpret the variation.

Nonetheless, studies of competition alone do not elucidate the effect of predation on organisms, or indeed the effect of predators on competitive interactions among the prey. Further, the relative importance of competition and predation in determining behavior, life history, population size and stability, and community structure is still controversial in ecology (Sih et al. 1985).

Predation and Competition

Examples of predator-mediated coexistence are common, both interspecifically (Slobodkin 1964; Hairston et al. 1968; Paine 1971; Wells 1970) and in maintaining morphs within a species (Kettlewell 1955; Zaret 1972). Some studies report that in the presence of predation more species will coexist than in the absence of predation (Addicott 1974). Some studies have failed to demonstrate predator-mediated coexistence (Harper 1969; Addicott 1974). With the controversy of whether predators do or do not allow more species to coexist has come an increase in the number of experimental studies examining the role of predators in community structure. Such studies of necessity have involved closed systems (Caswell 1978) where both prey and predators are confined in time and space, and usually involved removals or additions.

With the advent of numerous experimental studies of predation, Sih et al. (1985) reviewed and statistically analyzed the results of 139 studies with 1,412 comparisons from twenty years in seven journals. They noted whether the experiment resulted in the expected response (higher density of predators resulted in a reduction in prey species diversity, total prey abundance, prey population size, individual prey fitness, prey growth rate, or prey feeding rate) or the unexpected response (the opposite of the former responses). The keystone predator effect (Paine 1971) is thus an unexpected effect. Most field experiments (86 percent) on predation were done in the temperate zone. In the marine (nonintertidal) field experiments they described, only 5 percent were secondary carnivores, 53 percent were primary carnivores, and 42 percent were herbivores; 30.5 percent of the predators were vertebrates.

Almost every experimental field study on predation (95 percent) showed some significant effect (Sih et al. 1985). Over 85 percent of the studies had at least one comparison where prey showed a large response to predator manipulation, and 40 percent showed at least one unexpected effect. Oddly enough, predation effects were similar regardless of latitudinal zone, system (marine or not), predator trophic level, or experiment type when examined by study. However, when analyzed by the total number of comparisons, marine systems showed larger effects of predator manipulation than other systems. Sih et al. (1985) suggested that this may relate to the lower structural heterogeneity of marine systems, since structural simplicity increases the effectiveness of predators. Nonetheless, virtually no studies involving predator manipulation involve vertebrate prey.

The type of effect, rather than its mere existence, varied by latitude, location, trophic level, and taxon. Unexpected effects were more common 1) in temperate rather than tropical regions, 2) in freshwater and intertidal rather than other marine or terrestrial systems, 3) for primary carnivores, and 4) for vertebrate predators rather than invertebrates (Sih et al. 1985). Factors that mediate the effect of predators are structural heterogeneity and environmental stress (Mercurio, Palmer, and Lowell 1985). Structural heterogeneity provides spatial refuges where prey can hide, reducing predator efficiency. In laboratory experiments predatory fish foraged less efficiently in structurally complex environments because escape by prey was facilitated by structure (Savino and Stein 1982; Cook and Streams 1984); and in nature shorebirds feeding on mudflats did not severely affect prey abundance in areas with high sand concentration (Quammen 1984). Nonetheless, in marine systems, some avian predators have been shown to influence attributes of intertidal communities such as distribution patterns, abundance, and the size of the component invertebrate species (Castilla 1981; Edwards, Conover, and Sutler 1982; Hockey and Branch 1983; Mercurio, Palmer, and Lowell 1985). Kelp gulls *(Larus dominicanus)* severely affect the population size and structure of limpets *(Nacella delesserti)* in the subantarctic (Branch 1985). Similarly Whoriskey and Fitzgerald (1985) showed that three species of birds affect prey populations. In a given environment vertebrate predators may be less affected by stress than invertebrate predators. Recently investigators have been examining the effect of predators on guild composition and competitive interactions of prey (Morin 1984a, 1984b, 1986). They find complex relationships between predators and prey, whereby predators increase prey diversity.

The above section clearly indicates that both competition and predation affect population size and community structure. Sih et al. (1985) reanalyzed Connell's (1983) data set, and compared it to theirs, and showed that predators have more frequent and stronger affects than competitors. However, for other marine studies (nonintertidal) the effect was reversed: only 49 percent of the studies showed a predator effect whereas 55 percent showed a competitor effect (Sih et al 1985).

For marine birds, the indirect effects of competition and predation are critical. Multispecies, indirect effects occur when changes in a third species mediate the effect of one species on a second species. Prey life-styles profoundly affected by predation or competition include habitat use, activity budgets, foraging methods, diet, mating system, and life history.

Mutualism and Commensalism

Interactions that are positive for one or both members of a species pair are largely ignored in ecological studies, particularly in experimental studies. This could be because of their relative rarity, relative unimportance, sporadic occurrence in time and space, or a combination thereof. Nonetheless, such interactions occur, allowing some species or groups to coexist for limited periods of time. For marine birds, coexistence with other marine organisms is often limited spatially and temporally; thus positive interactions of avian predators and other vertebrate predators are possible with respect to prey distribution and capture (see below). Such positive interactions could occur at one trophic level that would have negative interactions on still other species at the same trophic level (via competition) or on another trophic level (via predation). Thus some predatory fish and some seabirds might positively affect one another because the behavior of each makes prey more available to the other (see Safina and Burger, this volume), but their behavior might negatively affect foraging success of other predatory fish or seabirds.

Marine Studies

Considering the abundance of observational and experimental field studies in marine and intertidal ecosystems, the lack of studies dealing with vertebrates either as competitors or predators is noteworthy. Even multidisciplinary studies of marine ecosystems have emphasized lower trophic levels and relatively small organisms (Schneider, Hunt, and Harrison 1986). Nonetheless, the competition-predation framework for analyzing interspecific interactions is usefully applied to vertebrates, if only on an observational basis. The lack of observational and experimental studies on marine vertebrates is partially a result of the unique attributes of aquatic systems, and of large organisms difficult to observe, confine, or manipulate. Since theories of causality in ecology seldom lend themselves to analyses by formal methods of hypothesis testing where the critical tests of alternative hypotheses can be performed, it is appropriate to be skeptical of hypothetico-deductive methods (Quinn and Dunham 1983). Nonetheless it is essential to begin to examine the complex marine ecosystems.

■
The Marine Interface: Open Versus Closed Systems

In ecological thought, the central problem of the coexistence of species has suffered from a number of biases. First, investigations have focused on specific ecosystems (intertidal, marine, lotic, lentic, terrestrial), ignoring the interfaces between systems. Second, studies examining the mechanisms of coexistence (e.g., predation, competition) have concentrated on species that dwell within one system at all stages in their life cycle (or at least for the stage being examined). This is particularly true for experimental rather than observational studies on competition and predation (Connell 1983; Sih et al. 1985). Third, most experimental studies on predation and competition do not examine birds or mammals, species that may be highly mobile during all or part of their life. The mobility of birds and mammals may result in their moving between ecosystems, living in different ecosystems for part of the time (either daily or seasonally), or indeed living in the interface between systems.

Until recently, the species interactions and coexistence mechanisms of marine birds in marine habitats have largely been ignored. Biologists have generally described the behavior of marine birds that were feeding, loafing, migrating, or breeding without reference to the marine environment or to the other organisms within it (Brown 1980). It is simply easier to observe birds from a ship deck with little reference to what is happening at or below the water's surface, although anecdotal references to marine birds feeding with marine mammals are legion.

The habitat of marine birds at sea is best considered as an interface zone extending from several meters above to several meters below the sea surface (the actual height and depth depending on the species of marine bird). Most species of marine organisms are limited to the sea itself, and the sea surface and shore act as a barrier. Animals living below the surface remain there except for unusual circumstances. Marine birds thus differ from other marine organisms in 1) spending most of their time on or above the sea in an air medium, 2) moving into and out of the water frequently, 3) being restricted to only a few centimeters to several meters of the top oceanic waters (for most species), and 4) being adapted for flight in air and thus usually being less mobile in water than they might otherwise be. Although birds move from an air to a water medium, other marine organisms are generally restricted to the water, creating an essential asymmetry. As a result of these factors, it is difficult to analyze the interactions of marine birds with other marine organisms as if they were all members of the same community at all times. Further, the average duration of interaction between marine birds and other marine species may be considerably less than the duration of interaction between marine fishes and marine mammals.

Considerations of the mechanisms of coexistence in classical ecological thought have usually involved closed systems where all members have equal access to the system. The asymmetries of the interface system where marine birds exist makes such an analysis difficult. Caswell (1978) discussed the theoretical approach of considering closed and open systems. A closed system is one with a closed, homogeneous volume of habitat with no migration, whereas an open system is a set of habitat cells coupled by migration. Although most ecological theory is based on closed systems, some laboratory studies have been based on open systems such as connecting cages (Huffaker 1958; Pimentel, Nagel, and Madden 1963). In population studies extinction can be countered by local recolonization. However, even with small numbers of simple components (habitat cells), the time required to reach equilibrium in open systems is long and may not be reached (Ashby 1960; Kauffman 1969). Thus, open systems are inherently unstable, although they are certainly important and relevant ecologically (Caswell 1978). Caswell (1978) developed a model for predator-induced coexistence of prey in open, nonequilibrium systems.

I suggest that marine interface systems are a special case of open, nonequilibrium systems in which the component cells are of three distinctly different types: 1) water surrounded by water, 2) water-air interface, and 3) air surrounded by air. Further, the marine interface, like some other systems, exhibits not only predator effects on prey and competition among prey (Schoener 1983; Connell 1983), but competition among predators.

The presence of three distinct types of cells provides a situation where local recolonization can occur only within some of the cells. For example, some organisms can live only in water and cannot enter the water-air interface, while others can live in the air and cannot enter the water. Still other organisms can live in one (either air or water) and can briefly pass through the air-water interface into the other cell type. In the latter case there is an asymmetry in that most aquatic marine animals can exist in air for only the briefest of time periods, while "air" organisms such as birds or marine mammals can remain in water for limited periods. Most marine birds, for example, usually are submerged for diving periods of less than 2.5 minutes (Dewar 1924; Dow 1964) although they may submerge to depths of 265 meters (Schorger 1947; Kooyman and Davis 1987; Jones and DeGange, this volume). The key characteristic is, of course, whether organisms must breathe in air or water. The asymmetry in allowable time in the "other" environment changes the nature of competition and predation. Animals living primarily in the air cells can be competitors and predators of organisms living primarily in the aquatic cells, but those in the aquatic cells are less likely to be predators of organisms remaining in the air cells.

Evolutionarily, it seems easier to evolve mechanisms for air-breathing animals (birds, some marine mammals) to live in the aquatic marine environment than for water-breathing animals (e.g., fish) to evolve mechanisms for existing totally in the air cells of the system. For example, pinnipeds move freely in all three cell types of the marine interface habitat even though they do not move as freely in air and certainly cannot fly. Birds move less freely than marine mammals in that their time submerged in the water is less. But nonetheless they can remain in the water for long enough to swim to the depth of several meters.

Ecological theory has been extensively developed for describing competition among organisms for a resource, predation on prey, herbivore effects on competing plants (Lubchenco 1978), and predation effects on competition among prey (see Schoener 1983; Connell 1983; Sih et al. 1985). However, in the marine interface ecosystem a fifth type of interaction may be very important: direct competition among a large number of predators during feeding may affect the predators as well as prey (both in numbers and in competitive interactions with other prey). Thus, several avian species might compete with each other or with fish or marine mammals for the same resource (e.g., prey fish). The former type of interaction is discussed by Hulsman (in this volume), while the latter is discussed further by Au, and Safina and Burger (in this volume). Even humans, as surrogate marine mammals through the use of hooks, traps, and nets of various sorts, can exert considerable influence on species interactions (Furness, Hudson, and Enser, and Jones and DeGrange, this volume).

The relative strengths of the effects of competition vary with the effectiveness of the predator and the allowable time (physiological constraints) in the prey environment. The asymmetry in allowable time in the prey's environment not only determines the presence and direction of competition and predation, but the relative importance of competition and predation to each organism. Competition among different orders is not usually studied (but see Levins 1979; Morin 1984a, 1984b). Yet in the marine interface these interactions may be prevalent and critical.

In the next section I discuss the types of interactions birds have with other marine animals in the marine interface. Marine interfaces include the interface between land and sea (at coastlines or the edges of islands) and between the air and sea over the open ocean. My discussion is not intended as an exhaustive list of examples, but as an introduction to the types of interactions possible.

■
Seabirds and the Marine Interface

Sea-Land Interface

Seabirds can coexist with other marine vetebrates on their breeding grounds or in the marine interface. Seabirds breeding in marine
environments either on islands, on coastal mainlands, or on barrier
beaches can interact with marine mammals such as seals or sea
lions. Interactions can take the form of competition for resting,
courting, or nesting sites, predation, or disturbance and depression
of breeding success (Warheit, Lindberg, and Boekelheide 1984).
    Breeding site competition for space between seabirds and marine
mammals is only just beginning to be investigated (see Pierotti, and
Warheit and Lindberg, this volume). Competition should be strongest on beaches where both seabirds and mammals breed, and where
suitable sites are limited (Figure 1.1). Mammals such as pinnipeds
are clearly constrained by their inability to move up onto such areas
(cliffs and jagged or steep rocks are impediments), and birds are

Figure 1.1. Brandt's cormorants *(Phalacrocorax penicillatus)* and California sea lions
(Zalophus californianus) breeding on the same rocky islands on Monterey Pennin-
sula, Calif.

constrained by the requirement for flat, flood- and predator-free nests sites.

Interference competition, or the effects of disturbance on reproductive success, have been demonstrated for black oystercatchers *(Haematopus bachmani)* along the Pacific coast (Warheit, Lindberg, and Boekelheide 1984). Interactions with pinnipeds result in abandoned nests and crushed eggs and chicks, thus lowering reproductive success. A lesser degree of interference competition occurs where seabirds and sea turtles use the same nesting beaches (Burger and Gochfeld, unpublished data). On Tryon and Masthead islands in the Great Barrier Reef (Australia), green sea turtles *(Chelonia mydas)* sometimes crawl over the nests of roseate terns *(Sterna dougalli)* and black-naped terns *(S. sumatrana)*, crushing the eggs or chicks. In three colonies of over three hundred tern nests we saw only three such nests destroyed.

Predation has seldom been documented between birds and marine mammals on breeding sites, although on the Pribilof Islands gulls *(Larus* spp.) eat the placentas and injured pups of northern fur seals *(Callorhinus ursinus;* Burger, unpublished data). Aurioles and Llinas (1987) recently reported on experiments suggesting that western gulls *(Larus occidentalis)* can injure and kill sea lion *(Zalophus californianus)* pups.

The Air-Water Interface

Abiotic Factors

In the early years the distribution of marine birds was studied only with respect to geographical location (latitude, continental shelf) or temperature (Brown 1966; Cooke and Mills 1972; Summerhayes, Hofmeyr, and Rioux 1974; Harrison 1982). The pelagic distribution of marine birds has been found to be influenced by oceanographic convergence and frontal systems (Brown 1979; Griffiths, Siegfried, and Abrams 1982; Schneider 1982; Kinder et al. 1983; Haney 1985; Duffy et al. 1984; Haney and McGillivary 1985; Schneider, Harrison and Hunt 1987), sea ice and ice edges (Cline, Siniff, and Erickson, 1969; Ainley and Jacobs 1981; Fraser and Ainley 1986), cyclones (Blomqvist and Peterz 1984), salinity (Gould 1983), and wind (Ainley and Boekelheide 1983). More recently, authors have tried to assess the relative importance of all of these environmental factors (Abrams et al. 1981; Hunt et al. 1981; Hunt, Eppley, and Drury 1981; Schneider, Harrison, and Hunt; in press

Haney 1986a). Clearly seabirds concentrate at fronts that mark the boundary between water masses in areas of strong surface flow gradient, although some species such as fulmars *(Fulmarus glacialis)* and auklets *(Aethia pusilla, A. cristatella)* seem unaffected by fronts (Kinder et al. 1983). Similarly, birds concentrate at ice edges where there is overlap between the bird community associated with pack ice and that associated with open water (Ainley, Wood, and Sladen 1978; Ainley, O'Connor, and Boekelheide 1984; Zink 1981).

The distribution of marine birds, however, is clearly the result of complex interactions between abiotic elements (ice systems, frontal systems), food availability, and nesting sites (during the breeding season—see Ashmole 1971; Hunt et al. 1981; Haney 1986a, 1986b). Presumably the distribution of prey species is also dependent upon abiotic features of the oceans. The complex relationships between seabirds and prey will be discussed below.

Biotic Interactions of Seabirds in the Marine Interface

In the marine interface, seabirds can engage in a variety of interactions with other vertebrates including: 1) preying on them (e.g., prey fish), 2) competing with each other for food, 3) competing with other vertebrate (fish or mammal) predators for prey fish, and 4) deriving benefits from the feeding activities of other vertebrates such as mammals or predatory fish. All four types of interactions can theoretically affect the community dynamics by affecting the density or characteristics (such as size, location) of predator or prey. In this section I briefly discuss the four major types of interactions and their consequences for community dynamics.

Seabirds as Predators

One of the major factors affecting the marine distribution of seabirds is prey availability and abundance. One method of assessing the relationship of seabirds and their prey is to examine correlations between seabird populations and prey populations. Veit, Braun, and Nikula (1987) showed that seabird populations wintering off the New England coast increased as sand lance *(Ammodytes* spp.) populations increased. Another method of assessing the food of seabirds is by examining the food brought back to mates and chicks in breeding colonies (Duffy 1983; Harrison, Hida, and Seki 1983; Hatch 1984; Gaston and Noble 1985; Duffy and Jackson 1986; Harris and Wanless 1986). More recently investigators have attached recording devices to diving birds or marine mammals to assess the speed and distance traveled while foraging and to determine energy expenditure (Wilson, Grant, and Duffy 1986). These studies give an accurate and useful picture of what parents eat or feed chicks, but not of the influence of seabirds on their prey, or of competitors, or of the influence of prey movements and dynamics on seabird behavior. Even combining data on seabird distribution over the oceans with food samples (from chicks or collected specimens) only implies relationships between prey availability and density (Hunt et al. 1981; Schneider and Hunt 1982a; Vermeer, Fulton, and Sealy 1985; Springer et al. 1984). Such studies, however, indicate resource partitioning (implying competition; see above) and often imply alterations in the abundance of prey (Springer et al. 1984; see paper in Nettleship, Sanger and Springer 1984). The effect of seabirds on their prey species can be particularly devastating when seabirds feed on eggs of their prey (Briggs et al. 1984).

Few investigators have censused seabirds and their prey fish (but see Schaffner 1986). However, Schneider and Piatt (1986) ran transects along the east coast of Newfoundland censusing both seabirds and schooling fish. They found significant correlations between numbers of birds and schooling fish. Extended aggregations of seabirds were associated with extended aggregations of schooling fish. Similar results were reported by Safina and Burger (1985, this volume) for common terns *(Sterna hirundo)* feeding over bluefish *(Pomatomus saltatrix)*. Whoriskey and Fitzgerald (1985) showed that three species of birds removed 30 percent of the sticklebacks *(Gasterosteus* spp.) in salt marsh pools. Likewise, declines in kittiwakes *(Rissa* spp.) were attributed to declines in pollock *(Theragra chalcogramma)* (Springer et al. 1986). Prey detectability by birds clearly relates to prey density (Eriksson 1985).

Commercial fisheries attempt to use seabirds to assess the location of fish and have studied the numbers, age distribution, or health of commercial fish species (Erdman 1967; Wahl and Heinemann 1979; Furness 1982a; Hatch 1984). As well as assessing the effect of seabirds on their prey, these analyses examine competition between seabirds and fisheries. Harvest estimates have been used with Brown pelicans *(Pelecanus occidentalis)* (Anderson et al. 1980). In this case the pelicans are dependent on Northern anchovies *(Engraulis mordax)*, a species increasingly harvested commercially. The implication is that people and pelicans are competing for the anchovies (Anderson and Gress 1984). Quantitative data on fish abundance, pelican abundance, and the effect of pelicans on their prey is correlational. Their model of the relationship between pelican reproduction and prey population suggests that fisheries and pelicans compete directly. Similar comparisons of the sizes of populations of commercial fish and of the seabirds that prey on them have been performed for several geographical regions (see papers in Nettleship, Sanger, and Springer 1984).

Bioenergetic models have been used to estimate the quantity of pelagic fish eaten by seabirds. Furness and Cooper (1982) estimated that three species of seabirds in Saldanha Bay (South Africa) consumed a total of 16,500 tons of fish a year, representing 30 percent of the mean annual catch in the fishing grounds. Similarly, Furness (1982b) estimated that seabirds consumed 29 percent of pelagic fish production within a 45-kilometer radius of a Scottish seabird colony. Furness used the historical approach of comparing commercial catches since 1890 with changes in seabird populations. Most seabird species have increased since 1900, probably in response to increased prey availability (Furness 1982b). Others have suggested that the effect of commercial fisheries is less important (Bourne 1983). Nonetheless, these estimates relate to commercial fisheries, and not to the effect of seabirds on their prey.

Recently, carbon flux models have been developed to compare trophic levels of seabirds among local regions (Schneider and Hunt 1982b). Trophic transfer to subsurface-feeding seabirds differed little between regions whereas trophic transfer in surface-feeding seabirds in the outer shelf and slope waters was three times greater than in waters of the middle shelf (Schneider and Hunt 1982b). Similar comparisons among ocean regions in the Northern and Southern hemispheres are now being made (Hunt 1985), and will no doubt lead to significant understandings of biomass transfer and ecosystem dynamics. Both methods lead to an understanding of trophic dynamics but do not provide experimental evidence for specific effects of predator-prey interactions.

All of these studies indicate that seabirds may consume significant quantities of commercially important fish. But there are no experimental studies showing the effect of seabirds on prey abundance and availability.

Seabirds as Competitors
Both diet studies and distribution studies indicate that seabirds potentially compete with one another for food both inter- and intraspecifically (as discussed above). Yet there are few studies that examine the behavior of seabirds for competitive interactions on

the foraging grounds. Instances of interspecific piracy indicate direct competition for already acquired food (Hopkins and Wiley 1972), although most of these studies take place at colonies rather than at foraging grounds. In Peruvian waters several species of seabirds feed in upwellings of high productivity, and dominance among species may be important in structuring flocks, and thus in determining which species obtains food (Duffy 1983). Abrams and Griffiths (1981), however, did examine the ecological structure of pelagic seabirds in the Benguela Current off the coast of southern Africa. They report trophic guilds of seabirds, and one guild dominates the assemblage at any one location. They believe that intraguild competition is reduced by interspecific differences in foraging behavior and temporal separation. Similarly, Duffy (1986) compared the foraging behavior of common and roseate terns on Long Island of the eastern United States and reported that the species appeared to partition food on the basis of patchiness. Common terns were more successful in large groups over large prey patches, and roseate terns were more successful in small groups over more dispersed prey. Further work on these two species on Long Island indicates that roseate terns are much more specialized and more restricted in their use of foraging habitats (Safina 1985).

Although such studies indicate that seabirds often use the same resources, it remains to be shown that such resources are in short supply, or that particular avian species suffer because of the consequences of using the same resources as other species. Such a conclusion is implied by current studies, but not shown by quantitative data.

Seabirds as Competitors with Other Vertebrates

One intriguing aspect of the marine interface is that large numbers of high trophic level predators congregate while feeding, and thus avian predators can compete directly with other predatory vertebrates. Predatory fish such as bluefish (Safina and Burger 1985), numerous scombrids (tunas *[Katsuwonus pelamis, Thunnus albacares]*; see Au and Pitman 1986), and various marine mammals (e.g., *Eschrichtius robustus*) feed on the same prey as seabirds (Harrison 1979). Although seabirds clearly derive benefits from the predatory fish and mammals (see below), it is also possible that they compete with them for prey fish. Seabirds also compete with pollock and whales *(Balaenoptera physalus)* for zooplankton (Springer and Roseneau 1985). Further, pollock is the largest contributor of biomass to diets of piscivorous seabirds on the Pribilofs (Hunt et al.

1981; Hunt, Eppley, and Drury 1981). Thus, the relationships are indeed complex. Several of these interactions are explored in this volume.

Seabird Commensalism and Mutualism
with Other Marine Vertebrates

Although seabirds may compete with other seabirds and with predatory marine mammals and fish for the same prey, seabirds clearly derive advantages from them including 1) fish school detection, 2) fish school compaction or fragmentation, and 3) fish availability. Certain species of seabirds arrive first over fish shoals, and may be used as guides by other seabirds and fisherman (Gochfeld and Burger 1982; Duffy 1983). Predatory fish and mammals often force prey fish toward the surface where prey respond by forming tight schools or by fragmenting and leaping (see Harrison and Schreiber 1985; Hall, Wardle, and MacLennan 1986). These two factors result in prey fish being more available (closer to the surface) and more easily detected by foraging seabirds. It is also likely that the predatory fish sometimes benefit because the rapid arrival of large numbers of certain seabirds (especially surface feeders) may occasionally prevent the prey fish from leaping above the water surface. Thus the prey fish are more available to the predatory fish (Colblentz 1985). Predatory fish pick off the individuals near the edge of the dense school and seabirds simply catch prey fish that are pressed up against the water surface. This phenomena occurs with laughing gulls *(Larus atricilla)* that form foraging flocks sitting above prey fish (Colblentz 1985).

Some associations between seabirds and other marine vertebrates have not been examined yet. For instance, on a trip off Monterey Bay (August 18, 1987), I noted a positive association ($\chi^2 = 5.5$, $P < 0.05$) between resting flocks of shearwaters *(Puffinus griseus)* and ocean sunfish *(Mola mola)*. Over 85 percent of the shearwater flocks had sunfish with them, and these tended to be the larger sunfish. There are numerous such relationships that need to be investigated.

■

Summary

Species coexist by a variety of mechanisms including competition, predation, and commensalism. Predator-mediated coexistence operates both inter- as well as intraspecifically. Most experimental studies of competition and predation have been in temperate regions and have largely ignored marine systems except for intertidal ecosystems. Most experimental studies of competition and predation have involved closed systems where the component parts remain within the system. Only recently have theoretical ecologists begun to develop models for open systems where all members do not have equal access to the system, and organisms move within the component cells. Such open systems are inherently unstable because of the time required to reach equilibrium.

Seabirds are members of an open system whereby they can be part of three cell types: water surrounded by water, water-air interface, and air surrounded by air. The interactions of seabirds with other marine vertebrates in the three cells depend on the behavior, ecology, and constraints of the other vertebrates as well as those of seabirds. Assymetries develop whereby the time each group can remain in each cell type varies depending on breathing physiology and anatomy.

Marine birds clearly interact with other marine vertebrates by competing with other seabirds, preying on prey organisms, competing with other vertebrates for food, and deriving benefits from the feeding activities of other vertebrates. In many cases our information on these interactions is ancedotal or descriptive rather than quantitative. Recently marine biologists have started to describe these interactions quantitatively, but experimental data are still lacking. The open nature of the marine interface and the size of the vertebrates involved makes experimental studies difficult. But questions of competition, predation, commensalism, mutualism, and species coexistence can nonetheless be profitably addressed in marine systems.

Acknowledgments

Over the years I have had many helpful discussions about seabird biology with Colin G. Beer, Richard G. B. Brown, Roger M. Evans, Michael Gochfeld, William A. Montevecchi, Ralph D. Morris, Ian C. T. Nisbet, John Ryder, Carl Safina, William Southern, and I thank them now. I also thank Michael Gochfeld and Carl Safina for critical comments on the manuscript, and Peter J. Morin for helpful discussions on the nature of competition.

References

Abrams, R. W. and A. M. Griffiths. 1981. Ecological structure of the pelagic seabird community in the Benguela current region. *Mar. Ecol. Prog. Ser.* 5:269–277.
Abrams, R. W., A. M. Griffiths, Y. Hajee, and E. Schoeppe. 1981. A computer-assisted

plotting program for analysing the dispersion of pelagic seabirds and environmental features. *Marine Ecology* 2:363–368.

Addicott, J. F. 1974. Predation and prey community structure: an experimental study of the effect of mosquito larvae on the protozoan communities of Pitcher Plants. *Ecology* 55:475–492.

Ainley, D. G., R. C. Wood, and W. J. L. Sladen. 1978. Bird life at Cape Crozier, Ross Island. *Wilson Bull.* 90:492–510.

Ainley, D. G. and S. S. Jacobs. 1981. Sea-bird affinities for ocean and ice boundaries in the Antarctic. *Deep Sea Research* 28:1173–1185.

Ainley, D. G. and R. J. Boekelheide. 1983. An ecological comparison of oceanic seabird communities of the South Pacific Ocean. *Studies in Avian Biology* 8: 2–23.

Ainley, D. G., E. F. O'Connor, and R. J. Boekelheide. 1984. The marine ecology of birds in the Ross Sea, Antarctica. *Ornithol. Monogr.* 32.

Anderson, D. W., F. Gress. K. F. Mais, and P. R. Kelly. 1980. Brown Pelicans as anchovy stock indicators and their relationships to commercial fishing. *CalCOFI Rep.* 21:54–61.

Anderson, D. W. and F. Gress. 1984. Brown Pelicans and the anchovy fishery off southern California. In D. N. Nettleship, G. A. Sangri, and P. F. Springer, eds., *Marine Birds: Their Feeding Ecology and Commercial Fisheries Relationships*, pp. 128–135. Dartmouth, Nova Scotia: Canadian Wildlife Service Special Publication.

Ashby, W. R. 1960. *Design for a Brain*. 2d ed. London: Chapman-Hall.

Ashmole, N. P. 1971. Seabird ecology and the marine environment. In D. S. Farner and J. R. King, eds. *Avian Biology*, 1:223–286. New York: Academic Press.

Au, D. W. K. and R. L. Pitman. 1986. Seabird interactions with dolphins and tuna in the eastern tropical Pacific. *Condor* 88:304–317.

Aurioles, D. and J. Llinas. 1987. Western Gulls as a possible predator of California Sea Lion pups. *Condor* 89:923–924.

Birch, L. C. 1957. The meanings of competition. *Amer. Natur.* 91:5–18.

Blomqvist, S. and M. Peterz. 1984. Cyclones and pelagic seabird movements. *Mar. Ecol. Prog. Ser.* 20:85–92.

Bourne, W. R. P. 1983. Feeding habitat of Black Skimmers wintering on the Florida gulf coast. *Wilson Bull.* 95:404–415.

Branch, G. M. 1985. The impact of predation by Kelp Gulls *Larus dominicanus* on the Subantarctic Limpet *Nacella delesserti*. *Polar Biol.* 4:171–177.

Brigggs, K. T., K. F. Dettman, D. B. Lewis, and W. B. Tyler. 1984. Phalarope feeding in relation to autumn upwelling off California. In D. N. Nettleship, G. A. Sanger, and P. F. Springer, eds., *Marine Birds: Their Feeding Ecology and Commercial Fisheries Relationships*, pp. 51–62. Dartmouth, Nova Scotia: Canadian Wildlife Service Special Publication.

Brown, R. G. B. 1966. Sea Birds in Newfoundland and Greenland waters, April–May 1966. *Canadian Field-Naturalist* 82:88–102.

Brown, R. G. B. 1979. Seabirds of the Senegal upwelling and adjacent waters. *Ibis* 121:283–292.

Brown, R. G. B. 1980. Seabirds as marine animals. In J. Burger, B. L. Olla, and H. E. Winn, eds., *Behavior of Marine Animals*, vol. 4: *Marine Birds*, pp. 1–40. New York: Plenum Press.

Castilla, J. C. 1981. Perspectivas de investigacion en estructura y dinamica de comunidades intermareales rocasas de Chile central, II: De predadores de alto nivel trofico. *Medio Ambiente* 5:190–215.

Caswell, H. 1978. Predator-mediated coexistence: A nonequilibrium model. *Amer. Natur.* 112:127–154.

Cline, D. R., D. B. Siniff, and A. W. Erickson. 1969. Summer birds of the pack ice in the Weddell Sea, Antarctica. *Auk* 86:701–716.

Colblentz, B. E. 1985. Mutualism between Laughing Gulls *Larus atricilla* and epipelagic fishes. *Cormorant* 13:61–64.

Connell, J. H. 1961. The influence of interspecific competition and other factors on the distribution of the barnacle *Chthamalus stellatus*. Ecology 42:710–723.

Connell, J. H. 1974. Field experiments in marine ecology. In R. Mariscal, ed., *Experimental Marine Biology*, pp. 21–54. New York: Academic Press.

Connell, J. H. 1975. Some mechanisms producing structure in natural communities: a model and evidence from field experiments. In M. Cody and J. Diamond, eds., *Ecology and Evolution of Communities*, pp. 460–490. Cambridge, Mass.: Harvard University Press.

Connell, J. 1980. Diversity and the coevolution of competitors, or the ghost of competition past. *Oikos* 35:131–138.

Connell, J. H. 1983. On the prevalence and relative importance of interspecific competition: evidence from field experiments. *Amer. Natur.* 122:661–696.

Cooke, F. and E. L. Mills. 1972. Summer distribution of pelagic birds off the coast of Argentina. *Ibis* 114:245–251.

Cook, W. L. and F. A. Streams. 1984. Fish predation on *Notonecta* (Hemiptera): relationship between prey risk and habitat utilization. *Oecologia* 64:177–183.

Dewar, J. M. 1924. *The Bird as a Diver*. London: Witherby.

Dow, D. D. 1964. Diving times of wintering water birds. *Auk* 81:556–558.

Duffy, D. C. 1983. The foraging ecology of Peruvian seabirds. *Auk* 100:800–810.

Duffy, D. C. 1986. Foraging at patches: interactions between Common and Roseate Terns. *Ornis Scand.* 17:47–52.

Duffy, D. C., A. Berruti, R. M. Randall, and J. Cooper. 1984. Effects of the 1982–3 warm water event on the breeding of South African Seabirds. *South African Journal of Science* 80:65–69.

Duffy, D. C. and S. Jackson. 1986. Diet studies of seabirds: a review of methods. *Colonial Waterbirds* 9:1–17.

Eadie, J. M. and A. Keast. 1982. Do Goldeneye and Perch compete for food? *Oecologia* 55:225–230.

Edwards, D. C., D. O. Conover, and F. Sutter. 1982. Mobile predators and the structure of marine intertidal communities. *Ecology* 63:1175–1180.

Erdman, D. S. 1967. Sea birds in relation to game fish schools off Puerto Rico and the Virgin Islands. *Caribbean Journal Science* 7:79–85.

Eriksson, M. O. G. 1979. Competition between freshwater fish and Goldeneyes *Bucephala clangula* (L.) for common prey. *Oecologia* 41:99–107.

Eriksson, M. O. G. 1985. Prey detectability for fish-eating birds in relation to fish density and water transparency. *Ornis Scand.* 16:1–7.

Fraser, W. R. and D. G. Ainley. 1986. Ice edges and seabird occurrence in Antarctica. *BioScience* 36:258–263.

Furness, R. W. 1982a. Competition between fisheries and seabird communities. In J. H. S. Blaxter, F. S. Russell, and M. Yonge, eds., *Advances in Marine Biology*, pp. 225–307. London: Academic Press.

Furness, R. W. 1982b. Seabird-fisheries relationships in the northeast Atlantic and North Sea. In D. N. Nettleship, G. A. Sangri, and P. F. Springer, eds. *Marine Birds:*

Their Feeding Ecology and Commercial Fisheries Relationships, pp. 162–169. Dartmouth, Nova Scotia: Canadian Wildlife Service Special Publication.

Furness, R. W. and J. Cooper. 1982. Interactions between breeding seabird and pelagic fish populations in the Southern Benguela Region. *Mar. Ecol. Prog. Ser.* 80:243–250.

Gaston, A. J. and D. G. Noble. 1985. The diet of Thick-billed Murres *(Uria lomvia)* in west Hudson Strait and northeast Hudson Bay. *Can. J. Zoo.* 63:1148–1160.

Gause, G. F. 1934. *The Struggle for Existence.* New York: Hafner.

Gochfeld, M. and J. Burger. 1982. Feeding enhancement by social attraction in the Sandwich Tern. *Behav. Ecol. Sociobiol.* 10:15–17.

Gould, P. J. 1983. Seabirds between Alaska and Hawaii. *Condor* 85:286–191.

Griffiths, A. M., W. R. Siegfried, and R. W. Abrams. 1982. Ecological structure of a pelagic seabird community in the southern ocean. *Polar Biol.* 1:39–46.

Hairston, N. G. 1980. The experimental test of an analysis of field distributions: competition in terrestrial salamanders. *Ecology* 61:817–826.

Hairston, N. G., F. E. Smith, and L. B. Slobodkin. 1960. Community structure, population control and competition. *Amer. Natur.* 94:421–425.

Hairston, N. G., J. D. Allan, R. K. Colwell, D. J. Futuyma, J. Howell, M. D. Lubin, J. Mathias, and J. H. Vandermeer. 1968. The relationship between species diversity and stability: An experimental approach with protozoa and bacteria. *Ecology* 49:1091–1101.

Hall, S. J., C. S. Wardle, and D. N. MacLennan. 1986. Predator evasion in a fish school: test of a model for the fountain effect. *Marine Biology* 91:143–148.

Haney, J. C. 1985. Wintering phalaropes off the southeastern United States: application of remote sensing imagery to seabird habitat analysis at oceanic fronts. *J. Field Ornithol.* 56:321–333.

Haney, J. C. 1986a. Seabird patchiness in tropical oceanic waters: the influence of sargassum "reefs." *Auk* 103:141–151.

Haney, J. C. 1986b. Seabird segregation at Gulf Stream frontal eddies. *Mar. Ecol. Prog. Ser.* 28:279–285.

Haney, J. C. and P. A. McGillivary. 1985. Midshelf fronts in the South Atlantic bight and their influence on seabird distribution and seasonal abundance. *Biological Oceanogr.* 3:401–430.

Harger, J. R. E. 1970a. Comparisons among growth characteristics of two species of sea mussel, *Mytilus edulis* and *Mytilus californianus. Veliger* 13:44–56.

Harger, J. R. E. 1970b. The effect of species composition on the survival of mixed populations of the sea mussels *Mytilus californianus* and *Mytilus edulis. Veliger* 13:147–152.

Harper, J. L. 1969. The role of predation in vegetational diversity. *Brookhaven Symp. Biol.* 22:48–62.

Harper, J. L. 1977. *Population Biology of Plants.* London: Academic Press.

Harris, M. P. and S. Wanless. 1986. The food of young Razorbills on the Isle of May and a comparison with that of young Guillemots and Puffins. *Ornis Scand.* 17:41–46.

Harrison, C. S. 1979. The association of marine birds and feeding Gray Whales. *Condor* 81:93–95.

Harrison, C. S. 1982. Spring distribution of marine birds in the Gulf of Alaska. *Condor* 84:245–254.

Harrison, C. S., T. S. Hida, and M. P. Seki. 1983. Hawaiian seabird feeding ecology. *Wildlife Monogr.* 85:1–71.

Harrison, C. S. and R. W. Schreiber. 1985. Mutualism between Laughing Gills *Larus atricilla* and epipelagic fishes. *Cormorant* 13:61–64.

Hatch, S. A. 1984. Nestling diet and feeding rates of Rhinoceros Auklets in Alaska. In D. N. Nettleship, G. A. Sangri, and P. E. Springer, eds., *Marine Birds: Their Feeding Ecology and Commercial Fisheries*, pp. 106–115. Dartmouth, Nova Scotia: Canadian Wildlife Service Special Publication.

Hockey, P. A. R. and G. M. Branch. 1983. Do oystercatchers influence limpet shell shape? *Veliger* 26:139–141.

Hopkins, C. D. and R. H. Wiley. 1972. Food parasitism and competition in two terns. *Auk* 89:583–594.

Huffaker, C. B. 1958. Experimental studies on predation: dispersion factors and predator-prey oscillations. *Hilgardia* 17:343–383.

Hunt, G. L., Jr. 1981. The Eastern Bering Sea shelf: Oceanography and resources. In D. W. Hood and J. A. Calder, eds., *The Eastern Bering Sea Shelf Oceanography and Resources*, 2:629–647. Washington, D.C.: NOAA.

Hunt, G. L., Jr. 1985. A preliminary comparison of marine bird biomass and food consumption between the southeastern Bering Sea and parts of the southern ocean. In W. R. Siegfried, P. R. Condy, and R. M. Lewis, eds., *Antarctic Nutrient Cycles and Food Webs*, pp. 487–492. Berlin and Heidelberg: Springer-Verlag.

Hunt, G. L., Jr., P. J. Gould, D. J. Forsell, and H. Peterson, Jr. 1981. Pelagic distribution of marine birds in the Eastern Bering Sea. In D. W. Hood and J. A. Calder, eds., *The Eastern Bering Sea Shelf Oceanography and Resources*, 2:649–687. Washington, D.C.: NOAA.

Hunt, G. L., Jr., Z. Eppley, and W. H. Drury. 1981. Breeding distribution and reproductive biology of marine birds in the Eastern Bering Sea. In D. W. Hood and J. A. Calder, eds., *The Eastern Bering Sea Shelf Oceanography and Resources*, 2:689–718. Washington, D.C.: National Oceanic and Atmospheric Administration.

Hutchinson, G. E. 1959. Homage to Santa Rosalia, or why are there so many kinds of animals. *Amer. Natur.* 93:145–159.

Kauffman, S. 1969. Metabolic stability and epigenesis in randomly constructed genetic nets. *J. Theoret. Biol.* 2:437–467.

Kettlewell, H. B. D. 1955. Selection experiments on industrial melanism in the Lepidoptera. *Heredity* 9:323–342.

Kinder, T. H., G. L. Hunt, Jr., D. Schneider, and J. D. Schumacher. 1983. Correlations between seabirds and oceanic fronts around the Pribilof Islands, Alaska. *Estuarine, Coastal, and Shelf Science* 16:309–319.

Kooyman, G. L., and R. W. Davis. 1987. Diving behavior and performance, with special reference to penguins. In J. P. Croxall, ed., *Seabird Feeding Ecology and Role in Marine Ecosystems*, pp. 63–75. Cambridge: Cambridge University Press.

Levins, R. 1979. Asymmetric competition among distant taxa. *Amer. Zool.* 19:1097–1104.

Lotka, A. H. 1932. The growth of mixed populations: two species competing for a common food supply. *J. Wash. Acad. Sci.* 22:461–469.

Lubchenco, J. 1978. Plant species diversity in a marine intertidal community: importance of herbivore food preference and algal competitive abilities. *Amer. Natur.* 112:23–39.

MacArthur, R. H. 1958. Population ecology of some warblers of Northeastern coniferous forests. *Ecology* 39:599–619.

MacArthur, R. H. and R. Levins. 1967. The limiting similarity, convergence and divergence of coexisting species. *Amer. Natur.* 101:377–385.

May, R. M., J. R. Beddington, C. W. Clark, S. J. Holt, and R. M. Laws. 1979. Management of multispecies fisheries. *Science* 205:267–277.

Menge, B. A. and J. P. Sutherland. 1976. Species diversity gradients: synthesis of the roles of predation, competition and temporal heterogeneity. *Amer. Natur.* 110:351–369.

Mercurio, K. S., A. R. Palmer, and R. B. Lowell. 1985. Predator-mediated microhabitat partitioning by two species of visually cryptic, intertidal limpets. *Ecology* 66:1417–1425.

Morin, P. J. 1984a. The impact of fish exclusion on the abundance and species composition of larval odonates: results of short-term experiments in a North Carolina farm pond. *Ecology* 65:53–60.

Morin, P. J. 1984b. Odonate guild composition: experiments with colonization history and fish predation. *Ecology* 65:1866–73.

Morin, P. J. 1986. Interactions between intraspecific competition and predation in an amphibian predator-prey system. *Ecology* 67:713–720.

Neill, W. 1972. Effects of size-selective predation on community structure in laboratory aquatic microcosms. Ph.D. thesis, University of Texas.

Nettleship, D. N., G. A. Sanger, and P. F. Springer, eds. 1984. *Marine Birds: Their Feeding Ecology and Commerical Fisheries Relationships.* Dartmouth, Nova Scotia: Canadian Wildlife Service Special Publication.

Pacala, S. W. and J. Roughgarden. 1982. An experimental investigation of the relationship between resource partitioning and interspecific competition in two two-species insular *Anolis* lizard communities. *Science* 217:444–446.

Paine, R. T. 1966. Food web complexity and species diversity. *Amer. Natur.* 100:65–75.

Paine, R. T. 1971. A short-term experimental investigation of resource partitioning in a New Zealand rock intertidal habitat. *Ecology* 52:1096–1106.

Pimentel, D., W. P. Nagel, and J. L. Madden. 1963. Space-time structure of the environment and the survival of parasite-host systems. *Amer. Natur.* 97:1410–1467.

Quammen, M. L. 1084. Predation by shorebirds, fish, and crabs on invertebrates in intertidal mudflats: an experimetnal test. *Ecology* 65:529–537.

Quinn, J. F. and A. E. Dunham. 1983. On hypothesis testing in ecology and evolution. *Amer. Natur.* 122:602–617.

Roughgarden, J. 1983. Competition and theory in community ecology. *Amer. Natur.* 122:583–601.

Safina, C. 1985. A Roseate by any other name would still be *Sterna dougallii. Bird Watcher's Digest* 7:66–68.

Safina, C. and J. Burger. 1985. Common tern foraging: seasonal trends in prey fish densities and competition with Bluefish. *Ecology* 66:1457–1463.

Savino, J. F. and R. A. Stein. 1982. Predator-prey interaction between Largemouth Bass and Bluegills as influenced by simulated, submersed vegetation. *Trans. Am. Fish. Soc.* 111:255–266.

Schaffner, F. C. 1986. Trends in Elegant Tern and Northern Anchovy populations in California. *Condor* 88:347–354.

Schneider, D. 1982. Fronts and seabird aggregations in the Southeastern Bering Sea. *Mar. Ecol. Prog. Ser.* 10:101–103.

Schneider, D. and G. L. Hunt, Jr. 1982a. A comparison of seabird diets and foraging distribution around the Pribilof Islands, Alaska. In D. N. Nettleship, G. A. Sangri, and P. A. Springer, eds., *Marine Birds: Their Feeding Ecology and Commercial*

Fisheries Relationships, pp. 86–95. Dartmouth, Nova Scotia: Canadian Wildlife Service Special Publication.

Schneider, D. and G. L. Hunt. 1982b. Carbon flux to seabirds in waters with different mixing regimes in the southeastern Bering Sea. *Marine Biology* 67:337–344.

Schneider, D. C. and J. F. Piatt. 1986. Scale-dependent correlation of seabirds with schooling fish in a coastal ecosystem. *Mar. Ecol. Prog. Ser.* 32:237–246.

Schneider, D. C., G. L. Hunt, Jr., and N. M. Harrison. 1986. Mass and energy transfer to seabirds in the southeastern Bering Sea. *Continental Shelf Research* 5:241–257.

Schneider, D., N. M. Harrison, and George L. Hunt, Jr. 1987. Variation in the occurrence of marine birds at fronts in the Bering Sea. *Estuarine, Coastal, and Shelf Science* 25:135–141.

Schoener, T. W. 1974. Resource partitioning in ecological communities. *Science* 185:27–39.

Schoener, T. W. 1982. The controversy over interspecific competition. *Amer. Sci.* 70:586–595.

Schoener, T. W. 1983. Field experiments on interspecific competition. *Amer. Natur.* 122: 240–285.

Schorger, A. W. 1947. The deep diving of the Loon and Old-Squaw and its mechanism. *Wilson Bull.* 59:151–152.

Sih, A., P. Crowley, M. McPeek, J. Petranka, and K. Strohmeier. 1985. Predation, competition, and prey communities: a review of field experiments. *Ann. Rev. Ecol. Syst.* 16:269–311.

Slobodkin, L. B. 1964. Experimental populations of Hydrida. *J. Anim. Ecol.* 33:131–148.

Springer, A. M., D. G. Roseneau, E. C. Murphy, and M. I. Springer. 1984. Environmental controls of marine food webs: food habits of seabirds in the Eastern Chukchi Sea. *Can. J. Fish Aquat. Sci.* 41:1201–1215.

Springer, A. M. and D. G. Roseneau. 1985. Copepod-based food webs: auklets and oceanography in the Bering Sea. *Mar. Ecol. Prog. Ser.* 21:229–237.

Springer, A. M., D. G. Roseneau, D. S. Lloyd, C. P. McRoy, and E. C. Murphy. 1986. Seabird responses to fluctuating prey availability in the eastern Bering Sea. *Mar. Ecol. Prog. Ser.* 32:1–12.

Strong, D., Jr., L. Szyska, and D. Simberloff. 1979. Tests of community-wide character displacement against null hypotheses. *Evolution* 33:897–913.

Summerhayes, C. P., P. K. Hofmeyr, and R. H. Rioux. 1974. Seabirds off the southwestern coast of Africa. *Ostrich* 45:83–109.

Veit, R. R., B. M. Braun, and B. Nikula. 1987. The influence of population growth of Sand Lance upon wintering seabirds in southeastern New England. *Pacific Seabird Group Bull.* 14:38.

Vermeer, K., J. D. Fulton, and S. G. Sealy. 1985. Differential use of zooplankton prey by Ancient Murrelets and Cassin's Auklets in the Queen Charlotte Islands. *Journal of Plankton Research* 7:443–459.

Volterra, V. 1926. Variatzioni e fluttuazioni del numero d'individui in specie animali conviventi. *Mem. R. Accad. Linnei Ser.*, vol. 6, no 2.

Wahl, T. R. and D. Heinemann. 1979. Seabirds and fishing vessels: Co-occurrence and attraction. *Condor* 81:390–396.

Warheit, K. I., D. R. Lindberg, and R. J. Boekelheide. 1984. Pinniped disturbance lowers reproductive success of Black Oystercatcher *Haematopus bachmani* (Aves). *Mar. Ecol. Prog. Ser.* 17:101–104.

Wells, L. 1970. Effects of Alewife predation on zooplankton populations in Lake Michigan. *Limnol. Oceanogr.* 15:556–565.

Whoriskey, F. G. and G. L. Fitzgerald. 1985. The effects of bird predation on an estuarine stickleback (Pisces: Gasterosteidae) community. *Can. J. Zool.* 63:301–307.

Wiens, J. A. 1977. On competition and variable environments. *Amer. Sci.* 65:590–597.

Wilbur, H. M. 1972. Competition, predation, and the structure of the *Ambystoma-Rana sylvatica* community. *Ecology* 53:3–21.

Wilson, R. P., W. S. Grant, and D. C. Duffy. 1986. Recording devices on free-ranging marine animals: does measurement affect foraging performance? *Ecology* 67:1091–1093.

Zaret, T. 1972. Predators, invisible prey, and the nature of polymorphism in the Cladocera (class Crustacea). *Limnol. Oceanogr.* 17:171–184.

Zink, R. M. 1981. Observations of seabirds during a cruise from Ross Island to Anvers Island, Antarctica. *Wilson Bull.* 93:1–20.

■ Commensalism

2

■ Associations Between Marine Birds and Mammals in the Northwest Atlantic Ocean

Raymond Pierotti · *Department of Zoology, University of Wisconsin, Madison*

The northwest Atlantic Ocean is a relatively small marine province (Briggs 1974). It is a zone of colder water effectively bounded to the east and south by the Gulf Stream component of the warm, saline North Atlantic Current (Hachey 1961). The area I discuss in this paper runs inshore of the Gulf Stream form Cape Hatteras to Greenland, with a northern boundary arbitrarily designated by a line from the Strait of Belle Isle to Cape Farewell at the southern tip of Greenland. Included in this area are the Gulf of Maine, the Bay of Fundy, the Gulf of St. Lawrence, and the offshore banks surrounding Cape Cod, Nova Scotia, and Newfoundland. In this paper, I concentrate on those areas of which I have personal experience: the Gulf of Maine, the inner parts of the Scotian Shelf, and southern and eastern Newfoundland out to the Grand Banks.

Nineteen species of seabirds breed in this area, including three Procellariiforms, three Pelicaniforms, three gulls, three terns, and seven alcids (Brown et al. 1975). One other alcid, the great auk *(Pinguinus impennis)* became extinct around the beginning of the nineteenth century (Brown 1985). In addition, approximately twenty other species migrate to this area either north from the Antarctic or the tropical Atlantic during the southern winter, or south from the Arctic during the northern winter (table 2.1).

Twenty-two species of cetaceans are residents for most of the year in the northwest Atlantic (Leatherwood and Reeves 1983; Min-

Table 2.1. Seabirds of the northwest Atlantic.

Breeding Birds	Southern Migrants	Nearctic Migrants
Northern fulmar *(Fulmarus glacialis)*	Cory's shearwater[a] *(Puffinus diomedea)*	Common loon *(Gavia immer)*
Manx shearwater *(Puffinus puffinus)*	Greater shearwater[a] *(Puffinus gravis)*	Red-throated loon *(Gavia stellata)*
Leach's storm petrel *(Oceanodroma leucorhoa)*	Sooty shearwater[b] *(Puffinus griseus)*	Red phalarope *(Phalaropus fulicarius)*
Northern gannett *(Morus [Sula] bassana)*	Audubon's shearwater[a] *(Puffinus lherminieri)*	Northern phalarope *(Lobipes [Phalaropus] lobatus)*
Great cormorant *(Phalacrocorax carbo)*	Wilson's storm petrel[b] *(Oceanites oceanicus)*	Pomarine jaeger *(Stercorarius pomarinus)*
Double-crested cormorant *(Phalacrocorax auritus)*	South Polar skua[b] *(Catharacta maccormicki)*	Parasitic jaeger *(Stercorarius parasiticus)*
Great black-backed gull *(Larus marinus)*		Long-tailed jaeger *(Stercorarius longicaudus)*
Herring gull *(Larus argentatus)*		*Great skua* *(Catharacta skua)*
Laughing gull *(Larus atricilla)*		Glaucous gull *(Larus hyperboreus)*
Black-legged kittiwake *(Rissa tridactyla)*		Iceland gull *(Larus glaucoides)*
Common tern *(Sterna hirundo)*		Thayer's gull *(Larus thayeri)*
Arctic tern *(Sterna paradisaea)*		Sabine's gull *(Xema sabini)*
Roseate tern *(Sterna dougalli)*		Ivory gull *(Pagophila eburnea)*
Razorbill *(Alca torda)*		Bonaparte's gull *(Larus philadelphia)*
Common murre *(Uria aalge)*		
Thick-billed murre *(Uria lomvia)*		
Dovekie *(Alle alle)*		
Black guillemot *(Cepphus grylle)*		
Atlantic puffin *(Fratercula arctica)*		

[a] Migrant from northern subtropics.
[b] Migrant from Southern Hemisphere.
[c] Rare in northwest Atlantic.

asian et al. 1984; table 2.2). Some species are rare throughout the year, and others, e.g., sperm and humpback whales, migrate south across the Gulf Stream, and right whales migrate south inshore past Cape Hatteras to breed in warmer waters (Gaskin 1982). There are four species of pinnipeds that occur in this area. All are phocids ("true," earless seals in the family Phocidae). Two are year-round residents, the harbor, or common, seal *(Phoca vitulina)* and the gray seal *(Halichoerus grypus)*, and two, the harp seal *(P. [Pagophilus] groenlandicus)* and the hooded seal *(Cystophora cristata)*, migrate south out of the Arctic to the pack ice to give birth in late winter, and return north with the pack ice into the Arctic in spring and summer (King 1983; Pierotti and Pierotti 1980). Until the nineteenth century, the walrus *(Odobenus rosmarus)*, the bowhead whale, *(Balaena mysticetus)*, and the gray whale (Eschrictius spp.) were also residents of this area, but these species were extirpated by ivory hunters and whalers (Gaskin 1982; Busch 1985).

■

Oceanography of the Northwest Atlantic

The northwest Atlantic Ocean is one of the most productive marine environments that is not a major upwelling zone (Hachey 1961; Powers 1983). This productivity is the result of several prominent oceanographic features that promote extensive mixing of warm and cold waters. Predominant among these is the Gulf Stream, and western edge of the North Atlantic Gyre. This band of fast-flowing, warm, saline water flows northeastward from the Caribbean into the north Atlantic, and effectively acts as a biogeographic barrier to the cold water–adapted fishes and invertebrates that inhabit the northwestern Atlantic (Hachey 1961; Briggs 1974). The principal cold water feature of this area is the Labrador Current, an outflow of the Arctic Ocean Gyre, which brings cold nutrient-rich water southeastward from Davis Strait of the Gulf to St. Lawrence and the Grand Banks of Newfoundland. This cold nutrient-rich water mixes with warmer water masses near the Grand Banks. This mixing generates seasonal bursts of productivity that provide the basis of a food chain resulting in large fish populations (Lilly 1981).

Much of this mixed water continues to flow south and west along the Scotian Shelf and to George's Bank and the Gulf of Maine. Tidal mixing over offshore banks causes deep-flowing nutrient-rich water to rise as it flows across them and results in an increase in primary

Table 2.2. Cetacean species of the north Atlantic.

Mysticetes	*Odontocetes*
Balaenopteridae	Physeteridae
Blue whale *(Balaenoptera musculus)*	Sperm whale *(Physeter macrocephalus)*
Fin whale *(B. physalus)*	Dwarf sperm whale *(Kogia simus)*[a]
Sei whale *(B. borealis)*	Ziphiidae
Minke whale *(B. acutorostrata)*	Northern bottlenose whale *(Hyperoodon ampullatus)*
Humpback whale *(Megaptera novaeangliae)*	True's beaked whale *(Mesoplodon mirus)*[a]
Balaenidae	Sowerby's beaked whale *(M. bidens)*[a]
Right whale *(Eubalaena glacialis)*	Blainville's beaked whale *(M. densirostris)*[a]
Bowhead whale *(Balaena mysticetus)*[b]	Delphinidae
Escrichtiidae	Bottlenose dolphin *(Tursiops truncatus)*
Atlantic gray whale *(Eschrictius* spp.?)[b]	Common dolphin *(Delphinus delphis)*
(Scrag whale)[c]	White-beaked dolphin *(Lagenorhynchus albirostra)*
	Atlantic white-sided dolphin *(L. acutus)*
	Risso's dolphin *(Grampus griseus)*
	Orca (killer whale) *(Orcinus orca)*
	Long-finned pilot whale (Pothead) *(Globicephala malaena)*
	Monodontidae
	Beluga *(Delphinapteras leucas)*
	Narwhal *(Monodon monoceros)*[a]
	Phocoenidae
	Harbor porpoise *(Phocaena phocoena)*

[a] Rare in northwest Atlantic.
[b] Extinct.
[c] Scrag whales were probably Atlantic Gray Whales but no one knows. The Scrag Whale was often reported by the whaling industry in the seventeenth and eighteenth centuries but was never formally identified. As a result, it does not have a formal scientific name.

productivity (Cohen et al. and Flagg et al. in Powers 1983). Some small rises in sea floor elevation off large banks, i.e., Stellwagen Bank in the Gulf of Maine and the Southeast Shoal of the Newfoundland Grand Banks, generate internal waves in the water column that create alternating convergences and divergences (Haury, Briscoe, and Orr 1979). Such areas mechanically create high concentrations of phyto- and zooplankton (Haury, Briscoe, and Orr 1979; Whitehead and Glass 1985).

The concentrated zooplankton supports large populations of fishes, including capelin *(Mallotus villosus)* and sand lance *(Ammodytes americanus)*. Larger fishes, i.e., cod *(Gadus morhua)*, hake *(Merluccius* spp.), and mackerel *(Scombrus* spp.), and long-finned squid *(Illex illecebrosus)* also gather in these areas to feed on capelin and sand lance (Winters and Carscadden 1978; Lilly 1981). The smaller fish species, capelin and sand lance, occur in very dense concentrations, especially during the inshore spawning and feeding migrations of capelin (Jangaard 1974; Carscadden 1984). Sand lance are the most abundant species from Nantucket Shoals north through George's Bank and the Gulf of Maine, whereas capelin become the most abundant species on the Grand Banks and northward to Labrador and the Davis Strait (Leim and Scott 1966; Lilly 1981).

These small, abundant fishes are important dietary items, not only for the larger species of fishes and squids, but also for large numbers of marine birds and mammals that occur in the area from Cape Cod north to Labrador and Greenland during the spring and summer (Winters and Carscadden 1978; Bradstreet and Brown 1985). Marine birds and mammals exploit these aggregations during times of high nutritional need (late May through September). For seabirds this is the time when chicks hatch out and must be fed, and the calves of small odontocete cetaceans are born and must be nursed (Gaskin 1982). Also during this period baleen whales are recovering from a four- to five-month fast during which they migrated to and from breeding areas, and females with young calves are still lactating (Gaskin 1982; Winn and Winn 1985; author's observations). At the same time, pinnipeds are recovering from their short but intense reproductive periods in winter or early spring, and pinniped young of the year are learning to forage (Pierotti and Pierotti 1980).

The small fishes and squid that make up most of the diet for marine birds and mammals either die or migrate offshore into deeper waters in the fall (Leim and Scott 1966; O'Dor 1983). As a consequence, food abundance is low compared with earlier in the year (Templeman 1948; Pitt 1958; Jangaard 1974; Lilly 1981). Small cetaceans and birds that are well adapted to live in the open sea, i.e., alcids and kittiwakes, follow these migrations offshore (Brown 1980, 1985). During this time of year the weather in the northwest Atlantic is often stormy and the water temperature declines to less than 5°C in many areas (Hachey 1961). By December, sea ice forms throughout the Gulf of St. Lawrence and the Labrador Sea, reaching as far south as Bonavista Bay in Newfoundland (Hachey 1961; Brown et al. 1975).

During the period from November through March or April the marine birds and cetacean fauna of the northwest Atlantic are either well out at sea, or out of the area (Brown et al. 1975; Powers 1983; Brown 1985), and much of their behavior and ecology during this period is little known. Some marine mammals respond to the north Atlantic winter by either leaving the area or by undergoing prolonged fasts. At this time humpback and right whales migrate south into warmer waters to feed or breed, with 80–90 percent of the individuals of these species fasting from December until the following April (Gaskin 1982). Gray seals haul out on land or sea ice to breed, spending much of the period from December through February fasting (Boness and James 1979). Pagophilic seals (harp and hood), move into the northwest Atlantic with the sea ice, but feed little if at all from January through April (King 1983; Pierotti and Pierotti 1980, 1983). Harbor seals remain close to shore throughout the year, feeding on bottom-dwelling fishes that have relatively constant populations. During winter harbor seals spend most of their time in the water, where their thick fat keeps them warmer than they would be on land (Boulva and McLaren 1979; Pierotti and Pierotti 1980, 1983).

Large, squid-eating cetaceans, e.g., sperm whales and ziphiids, are probably unaffected by climate as long as ice is not present, since they typically feed below the thermocline in deep cold water (Gaskin 1982). Delphinids are probably also little affected by winter since they concentrate primarily on offshore banks where topographical upwelling occurs, or at the front edge of the Gulf Stream, feeding on small fishes and squid, and only move inshore during spawning migrations of their prey (Gaskin 1982; author's observations).

■

Marine Bird and Mammal Associations

Kinds of Associations

Many of the marine mammals present in the northwest Atlantic do not appear to interact with marine birds. These include the deep-diving odontocetes that feed almost exclusively on mesopelagic squid, i.e., sperm whales and ziphiids. Other species of cetacean, e.g., blue whales, have not been observed to associate with marine birds (Evans 1982; author's observations).

The remaining species of marine mammals in the northwest Atlantic interact with seabirds to varying degrees (table 2.3). There are five possible forms of association between marine birds and mammals. The first, most simple form (henceforth called Type A), involves birds and mammals that occur in close proximity to one another but do not appear to interact. These birds and mammals may be exploiting different prey. An example of Type A association is gannets and humpback whales, since humpbacks typically feed on small schooling fishes or swarming invertebrates (Watkins and Schevill 1979; Brown 1980; Whitehead 1981), whereas gannets typically exploit larger fishes, such as mackerel (Ashmole 1971; Nelson 1978), which are probably also preying on the small fishes or invertebrates (Lilly 1981).

The second form of association (Type B) involves cetaceans and birds that simply seem to be attracted to the same resource, but do not show any positive attraction to each other. Type B associations may include phalaropes, shearwaters, and humpback whales, or petrels and right and sei whales, which congregate to feed on large swarms of euphausiids or copepods (Brown, Barker, and Gaskin 1979; Watkins and Schevill 1979; author's data).

The third form (Type C) involves birds that appear to be actively drawn to marine mammals because the foraging activities of the mammals drive or otherwise force prey to the surface where birds have access to a resource that otherwise would be unavailable (Harrison 1979; Watkins and Schevill 1979; Whitehead 1981). In a few of these cases, there exists the possibility of genuinely mutualistic interactions between cetaceans and marine birds. Type C associations include sulids or terns foraging with dolphins (Brown 1980), gulls feeding with dolphins (Wursig and Wursig 1979, 1980) or otariid pinnipeds (Ryder 1957; Pierotti, this volume), or various species

Table 2.3. Seabird-cetacean associations in the northwest Atlantic.

	SEABIRD SPECIES																	
CETACEAN SPECIES	GS	CS	SS	MS	NF	WSP	LSP	NG	GBB	HG	LG	BLK	CAT	TBM	CM	RB	AP	BG
Fin whale	A,B	A,B	A,B	A,B	A,B	A,B	A,B	A,B	A,B	A,B	na	A,B†	na	B	B	B	B	na
Minke whale	A,B†	A,B	A,B†	A,B†	na	B†	B	A,B†	A,B†	A,B	na	A,B†	na†	A,B	A,B†	A	A,B†	na
Sei whale	B	B	B	na	B	B	B	na	B	B	na	na	na	na	na	na	na	na
Humpback whale	B,C†	B,C	B	B,C	B	A	A	na†	B,C	B,C	B,C	B,C†	B,C	B	B	B	B†	A
Right whale	B	B	B	B	B	B	B	na	A,B	A,B	na	na	na	B	na	na	na	na
Pilot whale	B,C†	B	B,C†	B,C	B†	A†	A†	A†	B,C†	B,C†	na	B,C	A,B	B	B	B	A,B	A
White-beaked dolphin	B,C	B	B,C	B,C	B,D	A	A	B†	B,C	B,C	na	B,C	na	B	B	B	B	A
White-sided dolphin	B,C	B	B,C	B,C	B,D†	A	A†	na†	B,C	B,C	A	B,C	B,C	na	na	na	na	A
Harbor porpoise	B	B	B	na	na	na	na	na†	B	B	B	na†	B†	na	na†	na†	na†	A†

NOTE: Association types A–E are as described in text.

Abbreviations: GS = greater shearwater; CS = Cory's shearwater; SS = sooty shearwater; MS = Manx shearwater; NF = northern fulmar; WSP = Wilson's storm petrel; LSP = Leach's storm petrel; NG = northern gannet; GBB = great black-backed gull; HG = herring gull; LG = laughing gull; BLK = black-legged kittiwake; CAT = common Arctic tern; TBM = thick-billed murre; CM = common murre; RB = razorbill; AP = Atlantic puffin; BG = black guillemot; na, not observed by author.

† Association observed in British Isles by Evans (1982).

of birds and mammals feeding with humpback whales (Whitehead 1981; author's observations; see below).

In addition, some birds, principally gulls, skuas, fulmars, and giant petrels *(Macronectes)* are attracted to marine mammals because they scavenge by-products of the mammals (Type D). Birds feed on scraps from orca kills, or cetacean feces (Rice 1968; Condy, Van Aarde, and Bester 1978; Whitehead 1981; author's observations). Phocid seals interact with gulls or skuas on land or ice when the birds are feeding on recently expelled placentae, dead pups, or fecal material (Pierotti, this volume).

Finally, some marine mammals are predators on birds, so they may be attracted to assemblages of birds, but the birds show avoidance of the predators (Type E). Orcas eat all species of diving marine birds, i.e., alcids, cormorants, penguins, and shearwaters (Rice 1968; Condy, Van Aarde, and Bester 1978). Harbor seals take cormorants and gulls as prey by grabbing them from under the surface (author's observations), and gray seals take shearwaters (and other species) in much the same way (McCanch 1981).

Temporal and Spatial Patterns

Assemblages of marine birds occur primarily around breeding colonies during the period May through August. Major seabird breeding colonies in the area discussed are listed in Brown et al. (1975) and include the Witless Bay Islands off the Avalon Peninsula; Cape St. Mary's; Funk and Baccalieu islands in Newfoundland; Bonaventure, Anticosti, the Magdalene Islands in the Gulf of St. Lawrence; and Kent and Machias Seal islands in the Bay of Fundy/ Gulf of Maine. Around breeding colonies species composition is largely determined by prey type and the species that breed on the colony in question. Aggregations also occur in areas of high fish or invertebrate productivity throughout the year, such as the Grand Banks, George's Bank, and Stellwagen Bank in the southern Gulf of Maine.

Cape Hatteras to the Gulf of Maine

Ten species of seabird are numerically dominant in this area throughout the year, although specific dominants vary from season to season (Powers 1983). These are (in taxonomic order): the northern fulmar, Cory's shearwater, greater shearwater, sooty shearwater, Wilson's storm petrel, northern gannet, red phalarope, herring gull, great black-backed gull, and black-legged kittiwake. Of these species only the herring and great black-backed gulls breed in this area (Drury 1973–74).

During the spring and summer, birds observed inshore (within 0–2 kilometers of land) are almost exclusively gulls and cormorants (Powers 1983; author's data). During the spring (March–May), fulmars are the most common species observed offshore (6.5 birds/sq km), making up 48 percent of the species observed (Powers 1983). Gulls are the next most abundant birds observed offshore (5.25/sq km), making up 38 percent of the total north of George's Bank and 45 percent of the total in the Mid-Atlantic Bight (Powers 1983). Herring gulls occur primarily from 1–5 kilometers offshore (over 90 percent of 3,500 observations), and great black-backed gulls occur further offshore (author's data). During this season humpback whales migrate north from their winter calving grounds and are found in large numbers on George's Bank and throughout the Gulf of Maine and the Scotian Shelf (Winn and Winn 1985; author's observations). Fin and minke whales have returned to the area from offshore breeding grounds (Gaskin 1982). Harbor porpoises and white-sided and white-beaked dolphins are also abundant in these areas at this time (Gaskin 1982; author's observations).

During the summer months (June–August) shearwaters (15.3/sq km) and storm petrels (10.3/sq km) are the most abundant seabirds from the Mid-Atlantic Bight through the Gulf of Maine (Powers 1983). These include Wilson's storm petrels, and Cory's, greater, and sooty shearwaters (Powers 1983; author's data). Cetaceans, including humpback, fin, minke, sei, and right whales, occur in Type A and B associations with these marine birds (Overholtz and Nicolas 1979). Pilot whales (potheads) move inshore following squid migrations, and white-sided, common, and white-beaked dolphins are feeding on squid, herring, and sand lance (author's observations). These smaller cetaceans occur in Type B and C associations with marine birds at this time. During this period herring gulls remain the dominant seabirds around Cape Cod and on Stellwagen Bank, where several dozen humpback whales feed throughout the summer (Watkins and Schevill 1979; author's data). Also in summer, euphausiids swarm in large numbers in the southern Bay of Fundy, and gulls, shearwaters, phalaropes, and humpback and fin whales gather to feed on these swarms in Type A and B associations (Brown, Barker, and Gaskin 1979; Brown 1980; Gaskin 1982).

During the fall (September–November), greater and sooty shearwaters and Wilson's storm petrels migrate south and are almost completely absent from the northwest Atlantic by December (Brown et al. 1975; Powers 1983). At this time gulls become the most common seabirds (15.5/sq km north and 4.7/sq km south of Cape Cod), with approximately a million herring and 500,000 great black-backed gulls in offshore waters from the Chesapeake Bay to Nova Scotia (Powers 1983). Many of these birds are the young of the year, which concentrate either around fishing vessels or around hump-back whales (see below; author's observations). Humpback whales are in this area in large numbers (2,000–3,000) at this time while migrating southward from summer grounds off Newfoundland, La-brador, and Greenland to winter calving areas on shallow banks from the Dominican Republic down the Antilles to Trinidad (Whitehead 1981; Winn and Winn 1985). Other whales also migrate out of the area to offshore calving grounds (Gaskin 1982).

During winter (December–February), gulls remain the dominant seabirds (12.9/sq km) from Cape Cod to Nova Scotia (Powers 1983), but more birds have moved south of Cape Cod (6.1/sq km). Fulmars move into this area, replacing the departed shearwaters and storm petrels, and some alcids have moved south from their summer breeding grounds (Powers 1983; Brown 1985). Although most ceta-ceans are gone during this period, around 15 percent (400–500) of the western Atlantic humpback whale populations remains in the southern Gulf of Maine and around Nantucket Shoals, feeding throughout the winter (Winn and Winn 1985). In addition, some right and fin whales also remain in the southern Gulf of Maine during this period (C. Mayo, personal communication).

Atlantic Canada

Numerically dominant seabirds in this area are alcids (common and thick-billed murres, dovekies, Atlantic puffins, razorbill and black guillemots), Leach's storm petrel, northern gannets, and black-legged kittiwakes (see Brown et al. 1975 for population sizes). In contrast to New England waters, all of these birds breed in this area except the dovekie, which breeds primarily in Greenland (Brown et al. 1975; Brown 1985).

In early spring most of these species are concentrated in areas of high food availability, i.e., the Scotian Shelf or the Grand Banks. Nearly all alcids (murres, razorbills, puffins, and dovekies) are dis-tributed well offshore and further south than they are during the breeding season, concentrating on offshore banks from Newfoundland to Cape Cod (Brown et al. 1975; Brown 1985). Dovekies, which breed in Greenland, are on the Grand Banks, with a few individuals seen as far south as George's Bank (Powers 1983).

During the early spring (March–April), harp and hooded seals give birth on the pack ice in the northern Gulf of St. Lawrence and off northeastern Newfoundland (Sergeant 1965, 1973). Although these pinnipeds do not feed during this period, their placentae provide a rich source of food for gulls (King 1983). In late spring, cetaceans, including fin, minke, and blue whales, and white-beaked and white-sided dolphins, migrate back into this area to feed on capelin, mackerel, herring, and cod.

During early summer, humpback whales migrate onto the Grand Banks, and pass through the Witless Bay area on their way north to Labrador (Whitehead 1981; author's observations; see table 2.4). Blue and fin whales are common around Mingan Island in the Gulf of St. Lawrence during August (R. Sears, personal communication). Minke whales are found locally from Cape Cod to Newfoundland. In July, pilot whales move inshore following squid migrations (table 2.4). White-beaked dolphins sometimes join pilot whales in feeding aggregations (author's observations). During this season, most birds feed near their breeding colonies, but some forage further offshore, since great black-backed gulls, kittiwakes, murres, and puffins have been observed feeding on the western edge of the Grand Banks in July. Puffins and murres in these aggregations were observed flying westward with fish in their bills, which suggests that these birds were from the Witless Bay seabird colonies (author's observations).

In fall, birds leave their breeding colonies and migrate southward into warmer waters (Brown et al. 1975; Powers 1983; Brown 1985). Dovekies, thick-billed and common murres, and puffins winter on the Grand Banks and the Scotian Shelf, along with a number of kittiwakes, gulls, and fulmars (Powers 1983; Brown 1985). Most gulls, kittiwakes, and fulmars, however, winter on George's Bank and the Gulf of Maine, as do razorbills (Powers 1983; Brown 1985). Gannets winter in this area, but many gannets move well south of Cape Cod and are found offshore from the Chesapeake Bay to Cape Hatteras (Powers 1983).

Formation and Structure of Assemblages

The formation and structure of marine bird foraging assemblages in the northwest Atlantic are quite similar to those described by

Table 2.4. Herring Gull diets and presence of cetaceans around Great Island, Witless Bay, Newfoundland.

	May	June				July		
	1–31	1–7	8–14	15–21	22–30	1–7	8–15	16–23
1977								
Herring gull diets								
Capelin in diet (%)	0	0	5.9	47.2	61.4	65.4	73.3	20.8
Squid in diet (%)	0	0	0	1.5	4.5	9.6	21.8	78.6
Humpback whales around island (#)	0	1	3	12	14	16	12	3
Pods of pilot whales (#)	0	0	0	0	1	1	2	4
1978								
Herring gull diets								
Capelin in diet (%)	0	0	6.8	57.6	73.3	71.2	68.3	24.6
Squid in diet (%)	0	0	0	2.4	8.2	16.1	23.0	72.5
Humpback whale around island (#)	0	0	3	14	14	15	11	5
Pods of pilot whales (#)	0	0	0	0	1	1	3	4

Hoffman, Heinemann, and Wiens (1981). There are small, short-duration (5–30 minute) flocks consisting of 10–1,000 individuals (Type I of Hoffman, Heinemann, and Wiens), larger flocks consisting of several thousand individuals that last for hours (Type II), and flocks that may last days or even weeks that occur in fixed localities where strong upwelling or downwelling occurs (Type III).

The best examples of Type III flocks in the northwest Atlantic occur in the Bay of Fundy off Brier Island, Nova Scotia (Brown, Barker, and Gaskin 1979), and around the entrance to Head Harbor, New Brunswick (Gaskin 1982). A Type III flock also occurs in June and July on the Southeast Shoal of the Newfoundland Grand Banks (Whitehead and Glass 1985; R. G. B. Brown, personal communication).

In the Bay of Fundy tidal rips produce local upwelling that results in surface swarms of calanoid copepods and euphausiids that attract schools of mackerel, herring, and squid (Brown 1980; Brown, Barker, and Gaskin 1979; Gaskin 1982). This attracts large numbers of birds, especially in late summer and early fall when hundreds of thousands of phalaropes migrate through the Bay of Fundy (Brown 1980; author's observations). Other species of birds present include shearwaters and herring and great black-backed gulls off Brier Island, and Bonaparte's, herring, and great black-backed gulls off Head Harbor.

On the Southeast Shoal, capelin spawn in large numbers and cetaceans (mostly fin and humpback whales) gather to feed. Birds present are shearwaters, alcids, and gulls, including kittiwakes (Whitehead and Glass 1985).

Type III flocks in the northwest Atlantic typically have a substructure consisting of several types of association between birds and mammals. On all days (n = 14) that I have observed Type III seabird flocks, cetaceans were present in the aggregations. For example, fin whales feeding on euphausiids participate in Type A associations (see previous section) with gulls, alcids, and kittiwakes, which are feeding on fishes also drawn to the aggregation (n = 6), and Type B associations with phalaropes, shearwaters, and storm petrels, which feed on the euphausiids with the whales (n = 5) (author's observations; see also Brown, Barker, and Gaskin 1979; Brown 1980). In contrast, humpback whales feeding on schooling fishes were observed in Type C associations with gulls and kittiwakes (n = 14), and Type A and B associations with other seabirds (n = 14) (author's observations; see also Whitehead 1981).

Type II flocks occur in areas where capelin concentrate inshore during spawning in Newfoundland (n = 63 observed flocks) (author's observations), and offshore where pelagic crustaceans or fishes sporadically concentrate as a result of local upwelling, e.g., on Stellwagen Bank (n = 28 flocks). Shearwaters and gulls are the only species that were observed in both of these types of aggregations. Herring gulls, great black-backed gulls, and black-legged kittiwakes were observed in all inshore flocks in Newfoundland, and the two gull species were also observed in all offshore flocks. Greater and Manx shearwaters were observed in the Newfoundland aggregations (n = 22), and greater and sooty shearwaters were observed offshore (n = 19). Other bird species observed exploiting the schools of capelin in Newfoundland were murres (n = 19), puffins (n = 32), and razorbills (n = 9). Other bird species observed offshore feeding on crustaceans or fish were fulmars (n = 17) and Leach's and Wilson's storm petrels (n = 14). Type II flocks typically were found in association with cetaceans, primarily humpback (n = 24), fin (n = 11), and right whales (n = 3), but minke (n = 7) and sei whales (n = 2) were also observed to participate in assemblages (author's observations; see also Evans 1982).

The presence of the whales may provide an indication to seabirds of the presence of large concentrations of zooplankton (Type B association), since the whales are likely to remain in the vicinity of large concentrations of food (Gaskin 1982). On the eastern edge of Brown's Bank off southern Nova Scotia right whales attended by 80–100 storm petrels, 50–75 shearwaters, 4–6 fulmars, and 7–10 great black-backed gulls were found in the same loran location five days apart (author's observations). Nonbreeding birds may maintain a close Type B association with such whales for several hours to days, especially in the northern Gulf of Maine and off Nova Scotia and Newfoundland where dense summer fog restricts visibility and would make finding whales or concentrations of food more difficult for the visually oriented birds.

Many type I assemblages consist exclusively of birds. Observations of 308 of these all-bird assemblages have shown that they are of two basic types (author's data). Most (75 percent; n = 232) are small aggregations (10–30 birds) that last only 5–10 minutes and consist of gulls, kittiwakes, and gannets or terns. Typically a gull (including kittiwakes), tern, or gannet will plunge dive, which attracts a few gulls to the area. If the plunge diver surfaces with a fish and files away, the gull(s) will often pursue and attempt to steal the fish. In many cases, however, the gulls will plunge dive (43.5 percent; n = 101). If prey is captured and gulls begin to circle, more gulls and plunge divers often join the group (18 percent; n = 42) (see also Hoffman, Heinemann, and Wiens 1981). If there are shearwaters in the area they will quickly join such a group and plunge dive from just above the surface (17 percent; n = 39). Cetaceans were not observed to participate in these short-term assemblages (author's observations).

Alcids were also not observed to join such assemblages, possibly because they are more susceptible to predation and kleptoparasitism than other birds (Hoffman, Heinemann, and Wiens 1981). Another possibility, however, is that alcids avoid plunge divers, especially gannets. On twelve occasions, I have observed gannets diving into feeding groups of murres and puffins. In ten of these cases, the alcids were scattered and the groups broke up immediately. I did not observe gannets to forage in close association with cetaceans (but see Evans 1982). This may be to the benefit of both species when humpback whales are involved because a 2.5-kilogram gannet with a 15-centimeter bill plunge diving into a surfacing whale could result in damage to both individuals.

The alternate form of Type I assemblage typically involved hundreds of individuals of several species of seabird, exploiting large schools of small fish or squid (n = 76). Such assemblages lasted for longer periods of time (mean = 28 ± 17 minutes; range = 12–52 minutes) and had a more consistent structure. Those assemblages that involved only birds (n = 12) were as described by Hoffman, Heinemann, and Wiens (1981), with alcids around the periphery and shearwaters and gulls in the center. The functional roles of the various species were similar to those in the northern Pacific (Hoffman et al. 1981).

Gulls and kittiwakes acted as *catalysts* (terms and descriptions are from Hoffman, Heinemann, and Wiens 1981) whose conspicuous plumage and foraging activities were used by other species as indicators of the presence of prey. As with short-term assemblages, long-term assemblages did not form unless gulls or kittiwakes were present (n = 76). These species also acted as *kleptoparasites* on occasion, but appeared to capture most of their own prey, since they were observed to pick fish or squid off the water surface or just below the surface. Alcids functioned as *divers* in all ten flocks where they were present. Alcids were distributed around the periphery of the assemblage, and appeared to keep the school of prey concentrated by surrounding and diving underneath the school. Shearwaters sometimes acted as *suppressors* by diving into the center of the prey aggregation and scattering it (n = 3). The most efficient suppressors were gannets whose arrival and plunge diving activities broke up any assemblage within 20–30 seconds (n = 5).

An important component of the remaining sixty-four assemblages was cetaceans. The dynamics of these interactions will be described in the next section.

Type I Interactions Between Seabirds and Cetaceans

Small odontocetes, i.e., white-sided dolphins, white-beaked dolphins, and pilot whales, are extremely efficient at concentrating schooling prey (Norris and Dohl 1980; Evans 1982; Gaskin 1982). I have observed these gregarious, social cetaceans to concentrate prey in two primary ways. First, white-sided dolphins (n = 2 aggregations) and pilot whales (n = 7) drove schools of prey ahead of them by swimming parallel to each other until they caught up to the school. They then swam around and under the school, surrounding it and driving it toward the surface (Norris and Dohl 1980; Wursig and Wursig 1980; author's observations). The second method involved small groups of 5–10 white-sided or white-lipped dolphins which converged toward a common point from different directions (n = 2). Evans (1982) has described similar behavior in white-sided, white-lipped, and common dolphins, with fish leaping out of the water or moving in a "frenzied" manner ahead of the dolphins. Evans (1982) also describes gannets as "converging on the area and then plunge-diving in the vicinity".

I have not observed gannets in such aggregations, but I have observed gulls and terns to behave in a manner similar to that described by Evans (1982) for gannets. Gulls and terns dove into the water 2–5 meters ahead of lines of delphinids (n = 46 observations). Gulls also dove into the center of delphinid aggregations (n = 31), and surfaced with small fish or squid. Gulls were also observed to pick up prey attempting to escape from the cetaceans within a few centimeters of the surface (n = 37; author's observations).

In some areas humpback whales may be even more important to gulls than delphinids in driving prey to the surface. These large mysticetes feed on small fish near the surface by blowing bubbles underwater to form either "bubble-nets" or "bubble-curtains" (Juracz and Juracz 1979; Hain et al. 1982). These bubbles apparently frighten fish, which attempt to escape by swimming away from the bubbles to the surface. The most common species of seabird found associating with humpback whales were herring gulls (in the Gulf of Maine) and herring gulls and kittiwakes (Witless Bay). During the fall, many adults and hundreds of juvenile herring gulls were observed associating with humpback whales on Stellwagen Bank in the southern Gulf of Maine (n = 28 whales, 300–400 gulls; author's observations). During the late spring and summer, adult herring gulls were observed in association with all humpback whales seen feeding on small fish from Cape Cod to the Avalon Peninsula of Newfoundland (n = 107 whales), and kittiwakes were also commonly found associated with humpback whales off the coasts of Newfoundland (n = 66 whales) (author's observations; see also Whitehead 1981). Other species of birds found associating with humpback whales were shearwaters (n = 23 whales) in the Gulf of Maine, and shearwaters, murres, and puffins (n = 47 whales) in Newfoundland (author's observations; see also Whitehead 1981).

The standard method of food search employed by adult herring gulls was to fly high (50–100 m) above the water (n = 298 birds). When high-flying gulls spotted foraging groups of seabirds they dove quickly to the aggregation and left as the group broke up (n = 31). These birds than flew up to 25–50 meters above the water and resumed searching. When adult gulls spotted swimming humpbacks (n = 23) (or a fishing boat with nets in the water; n = 9) they circled above and then descended rapidly and sat on the water near the whale(s). If the whale(s) moved away the gulls took off and followed them.

When whales dove the gulls sat on the water or circled above until the whale(s) surfaced and then flew quickly near them. If whales blew a bubble cloud or net, the gulls took off and circled over the cloud (n = 49 observations of bubble clouds). Gulls dove at fish in the cloud, with a marked increase in gull diving and plunging activity as the whale surfaced (figures 2.1–2.3). Gulls continually flew over swimming groups of whales (n = 22 groups of whales and 134 gulls), but ignored whale groups that had fed and were engaged in social activity at the surface (n = 5 groups of whales and 49 gulls; author's observations).

Gulls had higher rates of feeding success when foraging in association with cetaceans, especially humpback whales (table 2.5). This increase in foraging success is almost certainly due to the tendency of humpbacks and small odontocetes to concentrate prey and drive them to the surface. Gulls ignored swimming fin (n = 13) and minke (n = 7) whales even when these species swam through foraging aggregations. These species do not drive prey to the surface in the northwest Atlantic, although minke whales do drive prey (principally herring) to the surface and associate with glaucous-winged and western gulls in Puget Sound (Dorsey 1983; author's observations).

Seabirds were also observed associated with right, sei, and fin whales feeding on concentrations of copepods or euphausiids (in Type II and III feeding assemblages; see above). Birds in these aggregations did not show any attraction to the whales, however, and ignored them even when whales swam through clouds of prey. This apparent indifference probably results from the failure of skim-feeding whales to concentrate prey (author's observations). In dense aggregations of invertebrate prey the birds do not require any concentrations generated by cetaceans. In fact birds may be disturbed or driven off particularly rich patches by cetaceans that disperse these patches as they swim through them.

In contrast, cetaceans in Type I assemblages may actually have their own foraging efficiency enhanced by the activities of birds. Dolphins, pilot whales, and humpback whales all feed primarily on small fishes (e.g., capelin, sand lance, and herring) and squid in the northern Atlantic (Mitchell 1975; Watkins and Schevill 1979; Whitehead 1981; Gaskin 1982; Hain et al. 1982). These prey form large schools that occur as dense but widely scattered patches within certain areas, i.e., the Southeast Shoal of the Grand Banks, Stellwagen Bank, and Witless Bay. Cetaceans concentrate near these areas (Whitehead 1981; Whitehead and Glass 1985; author's observations), but still must locate these widely scattered patches in order to feed. It is possible that some cetaceans follow schools of prey about, but this would be inefficient either when prey schools dove or when whales were not feeding.

Other cetaceans appear to remain in areas such as tidal rips or other zones of local upwelling where prey are predictably available on a fine-grained basis (Brown, Barker, and Gaskin 1979; Whitehead and Glass 1985). A third alternative, which seems to be used by humpbacks on Stellwagen Bank and near the Witless Bay seabird islands off the Avalon Peninsula of Newfoundland, is to disperse within a local area, i.e., 20–25 square kilometers, and search. As soon as any individual whale (or pod of delphinids) locates prey and beings feeding, it quickly attracts a number of gulls (including kittiwakes) that circle over the feeding cetaceans. The presence of large numbers of circling gulls quickly attracts both other seabirds (shearwaters and/or alcids) and other whales which join the aggregation.

In Witless Bay the diets of herring gulls appear to be strongly

Figure 2.3. Herring gull descending into bubble cloud produced by humpback whale in an effort to capture sandlance.

Figure 2.1. Herring gulls feeding on sandlance driven to surface by humpback whale using bubble cloud feeding. Bubble cloud is visible as pattern on water in foreground. *(Facing page, top.)*

Figure 2.2. Herring gulls accompanying humpback whale and feeding on sandlance. *(Facing page, bottom.)*

influenced by the presence of whales (table 2.4). During May and early June, herring gulls feed primarily on intertidal organisms, garbage, or other seabirds (Pierotti and Annett 1987). Capelin occur in the waters off Witless Bay in late May or early June (Carscadden 1984; author's observations; J. F. Piatt personal communication). Humpback whales typically arrive in the area in mid-June, and maintain a population of around 10–15 individuals through mid-July (table 2.4; J. F. Piatt, personal communication). The occurrence of capelin in gull diets corresponds strongly with the presence of humpback whales (table 2.4).

When humpback whales move out of the area, herring gulls switch their foraging association to pilot whales, which specialize in feeding on squid (Gaskin 1982). The squid are inshore to feed on capelin (Winters and Carscadden 1978), and other species of seabirds, i.e., puffins and murres, continue to take capelin (as do some gulls) throughout July (Bradstreet and Brown 1985). Most gulls, however, took squid during this period (table 2.4), and gulls were observed to take squid driven to the surface by pilot whales (see above).

Aggregations form quickly (within 5 minutes) around gulls feeding over whales. On Stellwagen Bank on June 21, 1985, I observed a female humpback whale accompanied by her calf blowing bubble

Table 2.5. Prey capture success rates of herring gulls.

Prey Species	Number of Captures Attempted	Number of Captures Successful	Percent Successful
Only seabirds in assemblage			
Capelin[a]	78	56	71.8
Capelin	106	37	34.9
Capelin	96	29	30.2
Sand lance	23	3	13.0
Sand lance	19	4	21.1
Squid	27	4	14.8
Assemblage with cetaceans present[b]			
Capelin	136	69	50.7
Capelin	172	114	66.3
Capelin	113	83	73.5
Sand lance	83	68	81.9
Sand lance	64	39	60.9
Long-finned squid	102	57	55.9
Long-finned squid	82	43	52.4

NOTE: Rates are for gulls foraging at sea in association with other seabirds, and with cetaceans present in the foraging assemblage.
[a] Capelin spawning in shallow water (less than 1 m deep), just offshore.
[b] Difference between assemblages with birds only and with cetaceans significant at .01 level by Wilcoxon Signed Ranks test.

clouds to drive sand lance to the surface. Initially the two whales were accompanied by about 20 adult herring gulls, and no other whales or seabirds were visible within 500 meters of the ship. Within five minutes, however, about 50 other gulls arrived accompanied by two more humpbacks, a minke, and a fin whale. Within fifteen minutes another 100 gulls arrived, as did three more fin whales, three more humpbacks, and approximately 300 greater and sooty shearwaters. Individuals joining the aggregation came from all directions of the compass, suggesting that they were not in groups prior to locating the aggregation.

This aggregation constantly changed its location within 3–5 square kilometers over the next hour, but the pattern of change remained similar. Typically, one or more fin whales would swim through a school of fish and disperse it, which caused the humpbacks and the gulls accompanying them to move away from the immediate vicinity. Gulls followed the humpbacks with about twenty to twenty-five gulls associating with each whale (see figure 2.2). When any individual humpback dove and surfaced or blew a bubble net the gulls accompanying it would circle, which attracted other gulls and the humpbacks associated with them. The entire assemblage would reform under the gulls until fin whales arrived (typically within 5–7 minutes), broke up the school of fish, and began the process anew.

I have observed fifty-three assemblages involving gulls and humpback whales and all have had this same basic structure and dynamics. That is, gulls located a feeding whale and other gulls and whales moved quickly to the location of the feeding. From a vantage point on Great Island in Witless Bay, I have seen humpbacks (n = 21) head directly for milling groups of gulls and kittiwakes from as far as 3 kilometers away on clear days. Gulls from the Great Island colony were observed to fly directly to groups of birds circling over whales more than 5 kilometers away (n = 37 groups). In all cases gulls flew to and from the aggregation, with approximately equal numbers of birds remaining with humpbacks until the whales finished feeding.

It is likely that humpback whales broadcast food location calls that attract other whales under these conditions, for humpbacks produce sounds during feeding activities (Winn and Winn 1985). It is also likely that humpbacks (and other whales) use the presence of circling gulls to locate food. Four lines of evidence support this argument. First, gulls moved much more quickly to join assemblages than did whales, and whales appeared to move directly to the location of circling gulls rather than the feeding whales which were often 20–30 meters behind the gulls. Second, the highest rate of spyhopping behavior (raising the head out of the water and looking around the aerial environment; Madsen and Herman 1980; Evans 1982) observed in humpbacks occurred when whales were near or in feeding aggregations, but were not feeding themselves at the moment (table 2.6). This suggests that whales seeking feeding companions looked as well as listened. Third, humpback whales have been observed to approach fishing boats hauling nets (n = 6; A. Knowlton, personal communication). It has been suggested that humpbacks scavenge fish spilled during fishing operations, but it is more likely that the whales are attracted to the large numbers of gulls that often attend fishing boats during net hauls (author's observations). It is possible to draw humpback whales (and dolphins) to a boat by chumming for and attracting large numbers of gulls. Finally, it is well established that other seabirds are attracted to foraging gulls (including kittiwakes) that act as catalysts for the formation of feeding assemblages (Sealy 1973; Hoffman, Heinemann, and Wiens 1981). If seabirds are capable of learning to use gulls as indicators of prey, it would be surprising if marine mammals as behaviorally sophisticated as cetaceans did not also make use of such information (see also Madsen and Herman 1980; Evans 1982).

I wish to emphasize that I do not think that birds require whales to find or capture food, or vice versa. Gulls are capable of exploiting a wide range of food types (Pierotti and Annett 1987), but they are also highly opportunistic, and cetaceans driving prey to the surface present a superb opportunity for gulls (including kittiwakes) to have

Table 2.6. Rates of spyhopping behavior by humpback whales.

| | | | ACTIVITY | | |
	Active feeding	Intervals between active feeding bouts	Traveling Alone	Traveling With others	Interacting with other whales
Number of hours observed	9.3	4.1	8.3	13.1	4.7
Number of spyhops/hour	0.32	5.85	0.12	0.23	2.55

access to a food source not usually obtainable because of the limited plunge diving abilities of gulls (Ashmole 1971). For most of the evolutionary history of gulls, feeding cetaceans (and other marine mammals) were the nearest functional equivalent of fishing boats. Since whales must also search large areas for patchily distributed food, it would be to their benefit to learn any cues to aid in location of prey patches. The conspicuous circling and milling of white birds (including gulls, terns, kittiwakes, and gannets) engaged in feeding would be one of the best cues available (Madsen and Herman 1980; Evans 1982). As a result, the foraging associations between cetaceans and high-flying white seabirds could have evolved as a facultative mutualism from which both groups benefit (see also Simmons 1972 and Evans 1982 for similar arguments).

■

Summary

In the northwest Atlantic, there are approximately forty species of marine birds and twenty-five species of marine mammals. Various species from these two groups are found in proximity to each other in five basic situations: Type A) birds and mammals in same area, but exploiting different food types and not interacting, Type B) birds and mammals attracted to the same food source, but not interacting, Type C) birds actively attracted to foraging mammals, Type D) birds feeding on by-products of mammals, i.e., feces and scraps of food, and E) mammals preying on birds. Of these, Type B and C associations are the most important.

In the northwest Atlantic, Type B associations typically involve mysticete cetaceans and Procellariiformes and phalaropes, but may also involve alcids, gulls, gannets, and coromorants, depending on the location and duration of the feeding flock/association. In contrast, Type C associations typically involve either small cetaceans or humpback whales and gulls (including kittiwakes). In Type C associations, there appears to be a possible facultative mutualism, in which the birds benefit from prey concentrated or driven to the surface by cetaceans, and the cetaceans may benefit from using the presence of conspicuously colored birds as a cue to the location of patchily distributed prey concentrations.

Acknowledgments

I thank R. G. B. Brown for critical and helpful comments on the ms.

References

Ashmole, N. P. 1971. Seabird ecology and the marine environment. In D. S. Farner and J. R. King, eds., *Avian biology*, 1:223–286. New York: Academic Press.

Boness, D. J. and H. James. 1979. Reproductive behaviour of the grey seal *(Halichoerus grypus)* on Sable Island, Nova Scotia. *J. Zool., London* 188: 477–500.

Boulva, J. and I. A. McLaren. 1979. Biology of the harbor seal in eastern Canada. *Bull. Fish. Res. Bd., Canada* 200:1–24.

Bradstreet, M. W. and R. G. B. Brown. 1985. Feeding ecology of the Atlantic Alcidae. In D. N. Nettleship and T. R. Birkhead, Eds., *The Atlantic Alcidae*, pp. 264–318. New York: Academic Press.

Briggs, J. C. 1974. *Marine Zoogeography*. New York: McGraw Hill.

Brown, R. G. B. 1980. Seabirds as marine animals. In J. Burger, B. L. Olla, and H. E. Winn, eds., *Behavior of Marine Animals*, vol. 4: *Marine Birds*, pp. 1–39. New York: Plenum Press.

Brown, R. G. B. 1985. The Atlantic Alcidae at sea. In D. N. Nettleship and T. R. Birkhead, eds., *The Atlantic Alcidae*, pp. 384–427. New York: Academic Press.

Brown, R. G. B., D. N. Nettleship, P. Germain, C. E. Tull, and T. Davis. 1975. *Atlas of Eastern Canadian Seabirds*. Ottawa: Canadian Wildlife Service Publication.

Brown, R. G. B., S. P. Barker, and D. E. Gaskin. 1979. Daytime surface swarming by *Meganyctiphanes norvegica* off Brier Island, Bay of Fundy. *Can. J. Zool.* 57:2285–2291.

Busch, B. C. 1985. *The War Against the Seals*. Montreal: McGill University Press.

Carscadden, J. E. 1984. Capelin in the northwest Atlantic. In D. N. Nettleship, G. A. Sanger, and P. F. Springer, eds., *Marine Birds: Their Feeding Ecology and Commercial Fisheries Relationships*, pp. 170–183. Dartmouth, Nova Scotia: Canadian Wildlife Service Special Publication.

Condy, P. R., R. J. vanAarde, and M. N. Bester. 1978. The seasonal occurrence and behaviour of killer whales, *Orcinus orca*, at Marion Island. *J. Zool., London* 184:449–464.

Dorsey, E. M. 1983. Exclusive adjoining ranges in individually identified minke whales in Washington State. *Can J. Zool.* 51:174–181.

Drury, W. H. 1973–74. Population changes in New England seabirds. *Bird-Banding* 44:267–313, 45:1–15.

Evans, P. G. H. 1982. Associations between seabirds and cetaceans: a review. *Mammal Rev.* 12:187–206.

Gaskin, D. E. 1982. *The Ecology of Whales and Dolphins*. London: Heinemann Press.

Hachey, H. B. 1961. Oceanography and Canadian Atlantic Waters. *Bull. Fish. Res. Bd., Canada* 134:1–120.

Hain, J. H. W., G. R. Carter, S. D. Kraus, C. A. Mayo, and H. E. Winn. 1982. Feeding behavior of the humpback whale in the western north Atlantic. *Fish. Bull.* 80: 259–268.

Harrison, C. S. 1979. The association of marine birds and feeding gray whales. *Condor* 81:93–95.

Haury, L. R., M. G. Briscoe, and M. H. Orr. 1979. Tidally generated internal wave packets in Massachusetts Bay. *Nature* 278:312–317.

Hoffman, W., D. Heinemann, and J. A. Wiens. 1981. The ecology of seabird feeding flocks in Alaska. *Auk* 98:437–456.

Jangaard, P. M. 1974. The capelin *(Mallotus villosus)*: biology, distribution, exploitation, utilization, and composition. *Bull. Fish. Res. Bd., Canada* 186:1–70.

Juracz, C. and V. Juracz. 1979. Feeding modes of the humpback whale, *Megaptera novaengliae*, in southeast Alaska. *Scient. Repts. Whales Res. Inst. Tokyo* 31:69–83.

King, J. E. 1983. *Seals of the World.* Ithaca, N.Y.: Cornell University Press.

Leatherwood, S. and R. R. Reeves. 1983. *Sierra Club Handbook of Whales and Dolphins.* San Francisco: Sierra Club Books.

Leim, A. H. and W. B. Scott. 1966. The fishes of the Atlantic coast of Canada. *Bull. Fish. Res. Bd., Canada* 155:1–485.

Lilly, G. R. 1981. Influence of the Labrador Current on predation by cod on capelin and sandlance off eastern Newfoundland. *Northwest Atl. Fish Org. Sci. Counc. Stud.* 3:77–82.

MacFarland, W. N. and E. R. Loew. 1983. Wave-produced changes in underwater light and their relation to vision. *Env. Biol. Fishes* 8:173–184.

McCanch, N. V. 1981. Predation on Manx shearwaters by grey seals. *Br. Birds* 74:348.

Madsen, C. J. and L. M. Herman. 1980. Social and ecological correlates of cetacean vision and visual appearance. In L. M. Herman, ed., *Cetacean Behavior*, pp. 101–148. New York: Wiley.

Minasian, S. M., K. C. Balcomb, and L. Foster 1984. *The World's Whales.* Washington, D.C.: Smithsonian Books.

Mitchell, E. D. 1975. Trophic relationships and competition for food in northwest Atlantic whales. *Proc. Can Soc. Zool.* 1974:123–133.

Nelson, J. B. 1978. *The Gannet.* Berkhamstead, England: Poyser.

Norris, K. S. and T. P. Dohl. 1980. The structure and function of cetacean schools. In L. M. Herman, ed., *Cetacean Behavior*, pp. 211–262. New York: Wiley.

O'Dor, R. 1983. *Illex illecebrosus.* In P. R. Boyle, ed., *Cephalopod Life Cycles*, 1:175–200. New York: Academic Press.

Overholtz, W. J. and J. R. Nicholas. 1979. Apparent feeding by the fin whale and humpback whale on the American sand lance in the Northwest Atlantic. *Fish. Bull.* 77:285–287.

Pierotti, R. and C. A. Annett. 1987. Reproductive consequences of dietary specialization and switching in an ecological generalist. In A. C. Kamil, J. R. Krebs, and H. R. Pulliam, eds., *Foraging Behavior*, pp. 417–442. New York: Plenum Press.

Pierotti, R. and D. Pierotti. 1980. Effects of cold climate on the evolution of pinniped breeding systems. *Evolution* 34:494–507.

Pierotti, R. and D. Pierotti. 1983. Costs of thermoregulation in adult pinnipeds. *Evolution* 37:1087–1091.

Pitt, T. K. 1958. Distribution, spawning, and racial studies of the capelin, *Mallotus villosus*, in the offshore Newfoundland area. *J. Fish. Res. Bd., Canada* 15:275–293.

Powers, K. D. 1983. Pelagic distributions of marine birds off the northeastern U.S. National Oceanic and Atmospheric Administration, Technical Memorandum NMFS-F/NEC-27. 1–201.

Rice, D. W. 1968. Stomach contents and feeding behavior of killer whales in the eastern North Pacific. *Norsk Hvalfangsttid.* 57:35–38.

Ryder, R. A. 1957. Avian-pinniped feeding associations. *Condor* 59:68–69.

Sealy, S. G. 1973. Interspecific feeding assemblages of marine birds off British Columbia. *Auk* 90:796–802.

Sergeant, D. E. 1965. Migrations of harp seals in the northwest Atlantic. *J. Fish. Res. Bd., Canada* 22:433–464.

Sergeant, D. E. 1973. Feeding, growth, and productivity of northwest Atlantic harp seals. *J. Fish. Res. Bd. Canada* 30:17–29.

Simmons, K. E. L. 1972. Some adaptive features of seabird plumage types. *Br. Birds* 65:465–479, 510–521.

Templeman, W. 1948. The life history of the capelin, *Mallotus villosus*, in Newfoundland waters. *Res. Bull. Nfld. Govt. Lab.* 17.

Watkins, W. A. and W. E. Schevill. 1979. Aerial observations of feeding behavior in four baleen whales: *Eubalaena glacialis, Balaenoptera borealis, Megaptera novaeangliae*, and *Balaenoptera physalus. J. Mammal.* 60:155–163.

Whitehead, H. P. 1981. *The Behaviour and Ecology of the Humpback Whale in the Northwest Atlantic.* Ph.D. thesis, Cambridge University, England.

Whitehead, H. and C. Glass. 1985. The significance of the Southeast Shoal of the Grand Bank to humpback whales and other cetacean species. *Can. J. Zool.* 63:2617–2625.

Winn, L. K. and H. E. Winn. 1985. *Wings in the Sea.* Hanover, N.H.: University of New England Press.

Winters, G. H. and J. E. Carscadden. 1978. Review of capelin ecology and estimation of surplus yield from predator dynamics. *Int. Comm. Northwest Atl. Fish. Res. Bull.* 13:21–30.

Wursig, B. and M. Wursig. 1979. Behavior and ecology of the bottlenose dolphin in the South Atlantic. *Fish. Bull.* 77:399–412.

Wursig, B. and M. Wursig. 1980. Behavior and ecology of the dusky dophin in the South Atlantic. *Fish. Bull.* 77:871–890.

3

The Structure of Seabird Communities: An Example from Australian Waters

Kees Hulsman • *School of Australian Environmental Studies, Griffith University, Nathan, Queensland, Australia*

Competition for some resource is often invoked by ecologists to explain why assemblages of species that live in the same area partition resources. The manner in which species partition their resources and how effectively they utilize them affect community structure. Here I am using the term "community" in the sense that two or more species co-occur during at least part of the breeding season. Thus there is a spatial and temporal overlap in their use of an area.

In this essay I explore how the nutrient and energy requirements of each species and the pattern of food abundance affect community structure. It is in this context that I examine the possible role of competition for food between various species of seabirds and their relationship with predatory fish.

The essay is divided into three parts. The first deals with a general perspective of structure of seabird communities. This perspective is developed in terms of Perrins' model (1970) of the relationship between food supply and timing of breeding, four hypothetical cases of food abundance patterns, and the classification of seabirds as inshore, offshore, and pelagic feeders.

The second part focuses on the structure of seabird communities of the Great Barrier Reef. Factors affecting the structure of thirteen seabird communities are discussed. Here the seabird community at One Tree Reef is described in detail and the following aspects are covered: the temporal and spatial overlap in foraging areas, foraging methods (including dependency on predatory fish to make food available), food size and type, and time of breeding.

The third part draws together points of discussion and inferences from theoretical (part 1) and field (part 2) data.

■
The Structure of Seabird Communities

One of the factors that seems to play an important role in the actual assemblage of species in an area is the pattern of resource abundance (Hulsman 1980). Perrins (1970) developed a model that explains the association between food supply and timing of breeding. He used two hypothetical cases to illustrate the model. I examine the general principles of the model and later consider their implications for timing of breeding and hence community structure.

Basically the model states that the abundance of available food has to be sufficient to allow birds to meet their increased nutrient and energy demands associated with breeding. An important component of the system that this model does not consider is the role of courtship behavior in bringing birds into a suitable physiological state for breeding. Nevertheless, let us assume that if the food supply is adequate the birds will become physiologically ready to breed.

The food supply is sufficient for birds to meet their maintenance and usual nonbreeding activity costs before time b1 and after time b2 (figure 3.1). The food supply is sufficient for females to meet maintenance costs as well as build up nutrient and energy reserves to form and lay eggs after time b1 until time b2. The time of laying depends on how quickly and effectively a female can attain sufficient nutrients and energy to form and lay eggs; this is the prelaying period and is PL in the figure. There is sufficient food available between times c1 and c2 to raise young successfully. However, young do not necessarily hatch at time c1. There is a delay in the system, namely time PL, and then the clutch has to be incubated and hatched (I) before the young can grow to fledging (F). In some cases it is possible for adults to raise young and molt at the same time (i.e., time d1 to d2).

The Pattern of Resource Abundance

To illustrate how the pattern of resource abundance affects community structure in seabirds I consider four different patterns of food abundance.

In Region 1, the food supply varies very little during the course of the year and usually remains above the level at which breeding can be sustained, i.e., the critical level (figure 3.2a).

In Region 2, the food supply is above the critical level for seven months of the year but it does not become superabundant (figure 3.2b).

In Region 3, the food supply is above the critical level for four months of the year during which it can be very abundant (figure 3.2c).

In Region 4, the food supply is above the critical level for two to three months of the year, during which it may be superabundant (figure 3.2d).

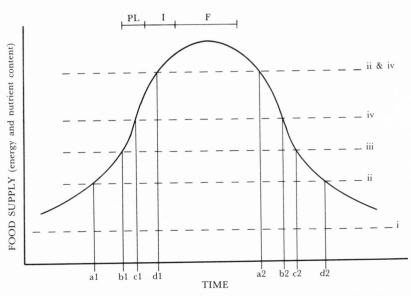

Figure 3.1. Diagram of the relationship between food supply and time of breeding in birds: i-maintenance and nonbreeding activity costs; ii-molt (a1 to a2); iii-forming eggs (b1 to b2); iv-rearing young (c1 to c2); ii + iv-molt and rearing young (d1 to d2). Adapted from Perrins 1970 (figure 1).

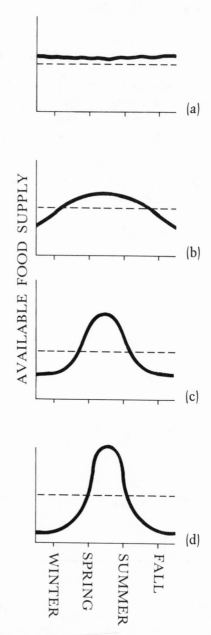

Figure 3.2. Pattern of resource abundance in four hypothetical regions. The broken line represents the critical level below which breeding cannot occur.

Seabird Utilization of Patterns of Food Abundance

If one applies Perrins' model to the pattern of food abundance in these four regions, one can specify some rules for the timing of breeding and molt in each region.

In Region 1, breeding can occur throughout the year. Here the actual time of seabird breeding may be influenced by the period required to complete molting because the food supply is not sufficiently abundant to allow adults to raise young and molt at the same time.

In Region 2, breeding may occur during seven months of the year but molt and breeding are partitioned as in Region 1.

In Region 3, breeding and molt may occur during the same period and are restricted to four months of the year.

In Region 4, breeding and molt occur during the same period, namely two or three months of the year.

Before considering the implications that these rules have on seabird communities in each of the four hypothetical regions, one must consider the general foraging and breeding characteristics of seabirds in a community.

Seabird Classification

Seabirds may be classified according to their feeding and breeding strategies. Categories often used by seabird researchers are inshore, offshore, and pelagic feeders (see Lack 1968; Diamond 1978). However, the criterion or criteria used to allocate a species to one of these three categories may vary according to the user (cf. Diamond 1978 and Hulsman 1980). Diamond classified species on the basis of the distance that they hunted from their colonies. In contrast, Hulsman classified them on the basis of the suite of characteristics that they exhibited. The basis for the latter approach is that foraging and breeding strategies are interdependent. Natural selection operates on them as an integrated package such that a change in one affects the other. Perhaps, more importantly, the extent of possible change in one characteristic is limited by others.

A gradient from inshore to pelagic feeders is probably a better description of seabird characteristics than three distinct categories. But the three categories can be identified along the gradient. Some species may show characteristics of more than one category and then we have the problem of which category to place them in. For example, the crested tern *(Sterna bergii)* sometimes hunts far from its colony and feeds its chick three to four very large fish per day. But otherwise it exhibits the characteristics of an inshore feeder such as fast growth rate and short fledging period. Thus it is easy to place this species on the gradient but not so easy to allocate it to one of the three categories. We could place it into the category with which it had most in common, but then some of the detail is lost. The simplified system is useful if it enables us to gain some insights into what affects the structure of seabird communities.

The characteristics of each of the three categories are as follows: the inshore feeder 1) has a clutch > 1; 2) has a short incubation period (< 28 days); 3) carries food in its bill to dependents; 4) hunts close to its colony; 5) feeds its young frequently (each feeding constitutes a small part of the young's daily intake); 6) has young that cannot tolerate going without food for long periods; and 7) has young that grow quickly and have short fledging periods (< 40 days).

The offshore feeder 1) has a clutch size of 1; 2) has a longer incubation period than an inshore feeder of comparable size; 3) carries food in its crop or stomach to dependents and therefore feeds them regurgitated food; 4) hunts out of sight of land including the colony; 5) feeds its young three to four times per day and each feed may provide a large amount of the chick's daily intake; 6) has young that can tolerate going without food for longer periods than those of inshore feeder; and 7) has young that have a slow growth rate and long fledging period (> 40 days).

The pelagic feeder 1) has a clutch size of 1; 2) has a long incubation period (> 28 days); 3) carries food in its crop or stomach to dependents and feeds them regurgitated food; 4) may range hundreds of kilometers from its colony in search of food; 5) feeds its young infrequently, e.g., once per day or once per week; 6) has young that can tolerate going without food for long periods; and 7) has young that have a slow growth rate and long fledging period (> 60 days).

Given these foraging types, I now will discuss the effect of the food supply (its pattern of abundance) on intra- and interspecific competition for food and on the structure of seabird communities.

Food Supply and Community Structure

In Region 1 there may be periods when larger numbers of seabirds breed than at other times, but breeding is possible at any time during the year. But as we have seen one of the implications of Perrins' model is that seabirds must partition the time when they breed and when they molt. If the population is in phase with its molt it may allow most of its members to breed at the same time. The advantages of breeding in large numbers coulbe 1) a reduction in the probability that a pair's young will be preyed upon, e.g., by predator swamping (Lack 1968; Nisbet 1975) or group defense (Kruuk 1964; Lemmetyinen 1971); and 2) an increase in the probability of locating prey, which are abundant but very patchily distributed, through the conspicuousness of feeding birds (see Phillips 1962; Lack 1968; Simmons 1972; Searley 1973) or information transfer (Ward and Zahavi 1973; Krebs 1978 in Burger 1981). The disadvantages of breeding in such large numbers could be 1) competition for nest sites and nest material (if used) (Lack 1968; Burger 1981); and 2) competition for access to food, causing some individuals to travel further from their colony to obtain sufficient food for their young (Ashmole and Ashmole 1967).

In this region, inshore feeders could utilize the food resources near the colony provided that they have some advantage over offshore and pelagic feeders. Prey tend to be more uniformly distributed and not as abundant near land as they are further from shore (Erwin 1977). Thus if inshore feeders can utilize dispersed prey more efficiently than offshore and pelagic feeders, they would have a place in the community (see the section on foraging methods, below; Hulsman 1978). Similarly, offshore feeders would have a place provided they have an advantage over inshore and pelagic feeders in the offshore zone. Pelagic feeders have the pelagic zone to themselves because the inshore and offshore feeders cannot utilize it.

Indeed the diversity of species in each of the three categories of seabirds forming a seabird community will depend on how, within each zone, the food resource is partitioned on the basis of access to prey (foraging methods and time of foraging), the abundance of prey species, and size class. In other words, the diversity of species is related to the number of habitats or temperature-salinity water types present (Ainley and Boekelheide 1983). Thus where food supply limits seabird species diversity then one would expect a greater diversity of inshore feeders, either offshore or pelagic feeders. The inshore zone consists of deep to very shallow water on the reef and therefore one would expect it to have more temperature-salinity water types than either of the other two zones. If this is true then it follows that where inshore feeders are less diverse than either of the other categories some factor other than food, e.g., the availability of suitable nesting areas, is responsible for determining community structure. I examine resource partitioning in more detail when dealing with the structure of seabird communities on the Great Barrier Reef (see below).

One outcome of the pattern of food abundance and energetic limitations as per Perrins' model could be a staggering in the times when different species breed. In view of the constraints placed on seabirds by this model, species that can collect sufficient food to meet their nutrient and energy requirements for breeding before the other species will breed first. A staggered breeding season can be a result of such a scenario. Competition for food will tend to enhance the staggering of the breeding season or lead to competitive exclusion.

If large concentrations of seabirds occur in the same local area (e.g., an island in an expanse of ocean) then competition for food could be severe. In these circumstances, interference competition for food as well as exploitive competition could regulate seabird numbers and largely shape the community structure. If the birds disperse over a large area during the nonbreeding season then competition would obviously be less severe at that time.

In Region 2, the duration of the potential breeding season is shorter than it is in Region 1. But the food supply is more abundant in Region 2 than it is in Region 1 during the breeding season. Therefore there is the potential for a greater variety of seabird species to coexist. The actual species diversity in each of the three seabird categories in Region 2 will depend on the characteristics of the food supply (e.g., the distribution and abundance of each size class of prey). As in Region 1, species in each of the three categories will have to have some advantage over species from either of the two other categories in their specific zone. Thus inshore feeders will have to have some advantage such as the ability to exploit dispersed prey more efficiently than can offshore and pelagic feeders and so on.

Competition for food is potentially less likely in Region 2 than in Region 1 during the breeding season because food is more abundant. However, the extent of competition for food in the two regions depends on how fully used the food supply is; in other words, the number of seabirds in relation to the carrying capacity of the environs. The degree to which the environs is saturated will depend on the survival of seabirds during the period when food is not so abundant.

In Region 3, the breeding season is four months long and although food is more abundant in the breeding season than it is in Regions 1 and 2, there is less time for staggered breeding. But then again, the potential for competition for food is less in Region 3 than in Regions 1 and 2 provided the numbers of each seabird species are limited by food or some other factor during the nonbreeding season. Species diversity in each of the three categories will depend on the same factors as in Regions 1 and 2, namely the ability to exploit prey in various zones but with less influence from competition for food and the availability of suitable nesting habitats.

In Region 4, food is superabundant during the short breeding season. There is no time for staggered breeding and it is unlikely that competition for food will occur unless it is interference competition. However, it is unlikely that food is the factor that limits species diversity in each of the three categories of seabirds during the breeding season. What occurs during the nonbreeding season is probably more important. Competition for food is more likely to occur in overwintering areas of these birds than during the breeding season.

Resource Partitioning in Great Barrier Reef Waters

I now direct attention to seabird communities in Australian waters, especially those of the Great Barrier Reef. There are twenty-one species of breeding seabirds using islands and waters of the Great Barrier Reef (Kikkawa 1976). Subsets of these twenty-one species co-occur in the same locale. Some locales have seven species whereas others have two. That leads to the question of what factors affect 1) the number of species in a locale, and 2) the number of species in each category of foraging and breeding strategy?

Of the ten species breeding in the Capricornia Region (table 3.1) only eight are sufficiently common to breed on all islands; the common noddy (= Brown Noddy, *Anous stolidus*) and the lesser crested tern *(Sterna bengalensis)* are rare. So why do the eight common species not nest on each of the islands? The answer lies in the availability of suitable nesting habitats in most cases. For example, the black noddy (*A. minutus* = white-capped Noddy, *Schodde, et al. 1978)* only breeds on islands that have suitable trees in which to nest (see Dale et al. 1984). Wedge-tailed shearwaters *(Puffinus pacificus)* nest on all islands that have suitable substrate in which to borrow. One Tree Island consists of coral rubble and hence no shearwaters nest there despite their attempts to dig burrows. Heron Island has so few species because human habitation and use has rendered it unsuitable for the ground-nesting seabirds. The few pairs of black-naped tern *(S. sumatrana)* that nest there are on a wreck at the harbor entrance and not on the island itself. North West Island has no ground-nesting seabirds because of its feral cat population. Thus the presence of a suitable nesting area has a critical effect on which species use an island (Hulsman 1980).

Given the specific habitat requirements of seabirds it also follows that the actual number of species in the inshore, offshore, and pelagic categories may be determined by the presence of suitable habitats in which to nest. Notwithstanding that, I direct attention to the possible role of competition for food. One Tree Reef is in a region where the pattern of food abundance is as in Region 2 (see figure 3.3; Kinsey 1972; Hulsman 1980).

Table 3.1. Species composition on each island in the Capricornia region of the Great Barrier Reef.

Islands	Shearwater (P)	Brown Booby (P)	Black Noddy (O)	Common Noddy (O)	Bridled Tern (O)	Crested Tern (I)	Lesser Crested Tern (I)	Black-Naped Tern (I)	Roseate Tern (I)	Pelagic	Offshore	Inshore	Species Diversity Index
Masthead	x		x		x	x		x	x	1	2	3	0.268
One Tree			x		x	x	x	x	x	0	2	4	0.636
Hoskyn	x	x	x		x			x	x	2	2	2	0.188
Lady Musgrave	x		x		x			x	x	1	2	2	0.275
Lady Elliot	x			x	x	x		x		1	2	2	0.275
Wreck	x			x				x	x	1	1	2	0.090
Erskine	x		x		x			x		1	2	1	0.278
Fairfax	x	x	x		x					2	2	0	0.282
Wilson	x				x			x	x	1	1	1	0.019
Tryon	x				x				x	1	1	1	0.060
North Reef						x		x	x	0	0	3	0.086
Heron	x		x					x		1	1	1	0.186
North-West	x		x							1	1	0	0.238

Abbreviations: P = pelagic; O = offshore; I = inshore.

Temporal and Spatial Overlap in Foraging Behavior and Diet

In this section I will deal with the time at which each species forages, where it forages, the foraging methods used, prey size and type, and the time at which each species breeds.

Time of Foraging

Generally the times at which inshore feeders foraged were related to the stage of the tidal cycle (figure 3.4) because of the movement of prey across the reef crest increased their availability, whereas the increased water depth decreased availability of prey that remain close to the substrate (Hulsman, Langham, and Bluhdorn in press). I have used the time at which chicks were fed to provide information about when a species foraged.

Crested terns were extremely variable in the times at which they foraged. On some days they hunted mostly during the rising tide and on others, during low tide (Hulsman, Langham, and Bluhdorn in press). Thus they tend to be opportunistic in when they forage and overall appear to forage with equal frequency throughout the whole tidal cycle (figure 3.4).

Lesser crested terns foraged more often during the rising and

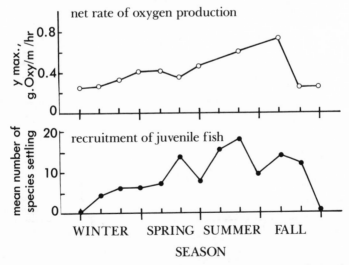

Figure 3.3. Annual pattern of primary production and recruitment of juvenile fishes at One Tree Reef. Adapted from Hulsman 1980; reproduced with permission of the Deutsche Ornithologen-Gesallschaft.

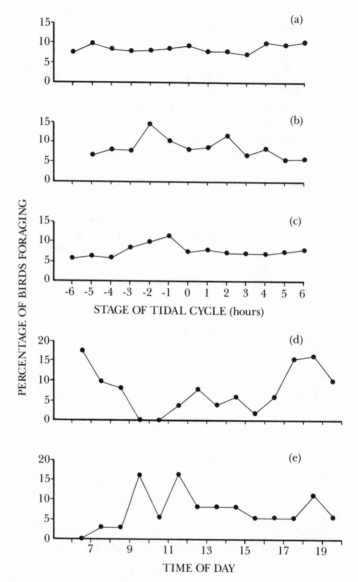

Figure 3.4 Frequency with which seabirds at One Tree Reef fed their young during the course of the tidal cycle or day: a-crested tern; b-lesser crested tern; c-black-naped tern; d-bridled tern; e-black noddy.

falling tide when their main prey, the atherinid *(Pranesus capricornensis)* crossed the reef crest (figure 3.4).

Black-naped terns foraged more often during the rising tide than at other times (figure 3.4). I have no data on the times at which roseate terns *(Sterna dougallii)* foraged.

Offshore feeders travel longer distances than inshore feeders and therefore one has to take into account traveling time to determine when they forage. Bridled terns *(Sterna anaethetus)* and black noddies did not hunt in relation to the stage of the tidal cycle. Bridled terns fed their young most often during the early morning and late afternoon (figure 3.4). Black noddies fed their young most often between mid and late morning, on one hand, and in the evening, on the other (figure 3.4). Both species hunted at sea where changes in water depth during the tidal cycle may not affect the availability of their prey as much as it does for inshore feeders because their prey occur near the surface. Furthermore, bridled terns often hunt singly and therefore do not readily drive their prey away from the surface (see Hulsman 1978). Noddies, on the other hand, hunt in dense flocks which may drive prey away from the surface; hence their dependence on predatory fish to make and keep prey available to them (Hulsman 1978). Thus bridled terns may hunt at specific times during the day whereas noddies hunt when predatory fish are active.

Foraging Zones

There is a significant association between species and foraging zone (χ^2 = 556.27, df 30, P < 0.001). This association can be clearly seen in figure 3.5. Crested and lesser crested terns differ greatly in where they foage. There seems to be a large overlap in where black noddies and crested, black-naped, roseate, and bridled terns hunt, although crested, black-naped, and roseate terns use the reef zones more than the other two species. The sea zone is a very large area and it may be oversimplifying the picture to lump all foraging activity in it into one zone. Therefore I will examine how the five species, which often use this zone, partition it.

Figure 3.6 clearly shows the difference between where the three inshore feeders (crested, black-naped, and roseate terns) and the two offshore feeders (bridled tern and black noddy) hunt in the sea zone. All three inshore feeders show a large overlap, but roseate terns venture farther from their colonies than either crested or black-naped terns (figure 3.6). There is a distinct difference in where most bridled terns and noddies hunt. Noddies tend to hunt closer to their

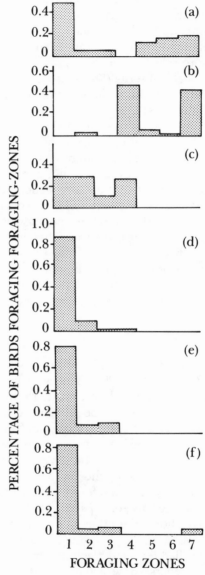

Figure 3.5. Frequency with which each species used each of the seven foraging zones: a-crested tern; b-lesser crested tern; c-black-naped tern; d-roseate tern; e-bridled tern; f-black noddy. 1-sea; 2-break-swell junction; 3-break; 4-small break; 5-reef crest; 6-reef flat; 7-lagoon.

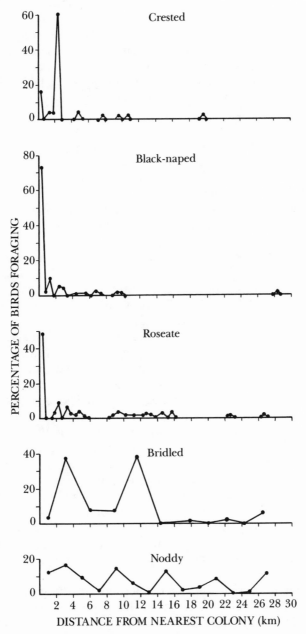

Figure 3.6. Frequency with which five species of tern forage at specific distances from their respective colonies.

colonies than bridled terns do. Most noddies were found hunting near reefs. The increase in noddies at 24 kilometers from their colony was near a reef without an island. The reason for this relationship between noddies and reefs may be because predatory fish activity near reefs is higher than in the open sea and noddies depend on predatory fish to make and keep their prey available (Hulsman 1978). This relationship will be elaborated on in the following section.

Foraging Methods

The mere fact that species utilize the same space at the same time or at different times does not provide sufficient information about possible competition for food between them. We must also consider how each species goes about catching its prey and of course the characteristics of its prey. First I will deal with how each species catches its prey because it partially explains the relationships between species in a community.

The color of a seabird's plumage is correlated with the foraging methods it uses. All of the *Sterna* species have white breast feathers. In contrast, the noddy has a completely dark plumage except for a white cap. According to Thayer-Craik's aggressive camouflage hypothesis a bird with a white underside can get closer to its prey before being detected than can one with dark plumage (Phillips 1962). Thus plumage coloration may explain why the *Sterna* species are able to air dive and plunge to the surface on dispersed prey and do not depend solely on predatory fish to make their prey available, whereas noddies hunt concentrated prey made available by predatory fish (Hulsman 1978; table 3.4).

All species plunge to the surface and five of the seven do it frequently (table 3.2). Plunging to the surface is when a bird dives into the water but does not submerge completely (see Ashmole and Ashmole 1967). The four inshore feeding terns use the air dive. Air diving is when a bird dives into the water and submerges completely, relying on its momentum to carry it to its victim (see Ashmole and Ashmole 1967). Terns plunge or dive toward their prey and invariably grasp them just behind the opercula. This is the best place to grasp a fish to minimize its chances of escaping a predator's grasp.

The four terns that are inshore feeders can hunt prey that are widely dispersed as well as those that are more concentrated in schools. They hunt in shallow water (e.g., small break, over patch reefs, etc.), but sometimes they form flocks in these areas. Such behavior does not provide much scope for interference competition between these species, at least not on the scale of that in noddy flocks, as discussed below.

The depth to which a tern dives depends on its momentum, the angle at which it enters the water, the bird's buoyancy, and water resistance. Hulsman (in prep.) presents a model that illustrates the interrelationships between these variables. For example, although crested terns dive from 9 meters and lesser crested terns from 10 meters, the former enter the water at right angles whereas the latter enter at more acute angles (Table 3.3). Therefore crested terns with their greater mass will dive deeper than lesser crested terns provided the greater bouyancy of the crested tern does not negate the effect of the angle of entry. Black-naped and roseate terns will not dive as deeply as the two large terns because of the former's much smaller mass and the fact that they dive from slightly lower heights (table 3.3). There is probably very little difference in the depths reached by black-naped and roseate terns because the slightly greater mass of the roseate tern may be counteracted by the slightly greater height and angle that black-naped terns use as well as the roseate tern's slightly greater bouyancy. Thus these inshore feeders will be vertically segregated.

The angle of entry may affect forgaging success. If a tern approaches a prey at an angle of less than 90° there is more chance for error in judging the prey's true position than if it approaches at an angle of 90° because of parallax error. This is important because

Table 3.2. Foraging methods used by six tern species nesting at One Tree Island.

| | FORAGING METHOD | | | | |
SPECIES	Air dip	Contact dip	Air dive	Plunge to surface	Hover
Crested	x		*	*	
Lesser crested			*	*	
Black-naped	x		*	*	
Roseate			*	*	
Bridled		*		x	
Black noddy	x	x		*	x

NOTE: Terminology from Ashmole (1971).
x = used
* = frequently used

terns do not swim underwater and are probably committed to the
trajectory determined by their angle of entry and the subsequent
deviation caused by their bouyancy. The slower a tern moves through
the detection field of the prey the more chance the latter has of
escaping. Thus plunging from a lower height may affect the velocity
attained and the likelihood of success.

Differences in heights and angles at which terns plunge to the
surface, in contrast to the air dive, make very little difference in the
depth reached; the bird remains at the surface. But these characteristics may affect foraging success because of the speed at which a
bird moves through the detection field and parallax error.

The offshore feeders use different foraging methods than those
used by the inshore feeders. Bridled terns usually contact dip and
therefore catch prey in the upper 4 centimeters of the water column
(Hulsman and Langham 1985). They tend to hunt dispersed prey
but occasionally join flocks of noddies. Therefore they do not rely
solely on predatory fish to make prey available (table 3.4).

Black noddies essentially drop from less than 1 meter onto their
prey at the surface (Hulsman and Langham 1985); for want of a
better term I have included this in the category of "plunge to the
surface." Noddies, unlike other species, rarely forage alone. They
form flocks ranging in size from 4 to more than 2,600 birds (mean
= 200, s.d. = 348, n = 296). There is enormous potential for
interference competition with other birds in the flock because the
noddies concentrate where the prey break the surface. The potential
is probably reduced somewhat by the organization within the flock.

Table 3.3. Air diving and plunging of four inshore-feeding terns.

		AIR DIVE		PLUNGE TO THE SURFACE	
SPECIES	Weight (g)	Height (m)	Angle (degrees)	Height (m)	Angle (degrees)
Crested	350	9 ± 1.5(17)	90 ± 0 (17)	5 ± 0.5(62)	72 ± 3.8(40)
Lesser crested	240	10 ± 0.6(60)	80 ± 5.6(62)	6 ± 1.1(33)	53 ± 8.2(48)
Black-naped	100	8 ± 0.4(35)	89 ± 1.3(35)	2 ± 0.2(96)	79 ± 3.9(67)
Roseate	120	7 ± 0.3(34)	87 ± 1.2(34)	5 ± 0.5(21)	83 ± 3.6(21)

NOTE: Heights and angles are ± a 95 percent confidence limit. If mean falls outside the 95
percent confidence limits of another sample then there is a significance between the means.
Sample sizes are given in parentheses.

Table 3.4. Percent occurrence of each species foraging solitary or in flocks and with one another in those flocks.

	SPECIES					
	Crested	Lesser crested	Black-naped	Roseate	Bridled	Black Noddy
Number of occurrences	328	1760	2935	1051	731	531
Percent occurrence feeding solitarily	98.2	99.2	95.9	96.3	98.9	0.0
Percent occurrence feeding in flocks	1.8	0.8	4.1	3.7	1.1	100.0
Percent occurrence over other predatory fish	1.6	0.3	1.6	1.1	0.1	20.5
Percent occurrence over tuna	0.2	0.5	2.5	2.6	1.0	79.5
Crested		0.6	1.5	0.6	0.3	2.6
Lesser crested			4.6	0.3	0.3	6.0
Black-naped				12.3	2.8	50.6
Roseate					1.2	20.9
Bridled						3.4

But chaos (disorganization) reigns supreme where bait fish, such as *Engraulis australis* and *Sprattus sprattus*, break the surface. Before dealing with the chaos in the flock I will consider the degree of orderliness or pattern in it.

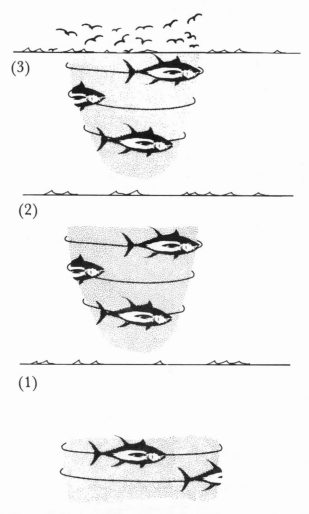

Figure 3.7. Diagram of how predatory fish herd bait fish to the surface: 1) predatory fish start circling a school of fish and compact it; 2) circling predatory fish lift the bait fish toward the surface; 3) bait fish are at the surface where terns may catch them. The next stage is when fish dart through the school feeding on bait fish which then escape by diving, and the process is repeated.

Predatory fish such as the bonito *(Euthunnus affinus)* and yellow-fin tuna *(Neothunnus macropterus)* drive bait fish to the surface. A school of tuna herd bait fish by circling them at various levels, compacting them and taking them to the surface where both the tuna and birds feed on them (B. Coates, personal communication; figure 3.7). There is a feeding frenzy at the surface, and tuna charge through the school, breaking it up. The bait fish escape the predators when there are too few tuna circling them to keep them at the surface. In the short time that they are at the surface, noddies feed (1.30 ± 0.66 min, n = 29). Then the fish dive and become unavailable for a short period (1.76 ± 1.89 min, n = 25) while tuna reherd them and bring them back to the surface. Meanwhile noddies follow the pattern depicted in figure 3.8. A noddy moves low over the

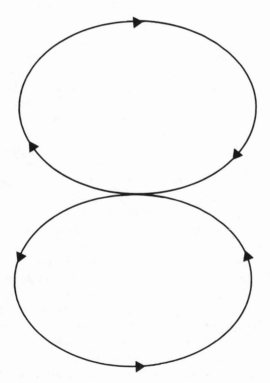

Figure 3.8. Overhead view of pattern followed by black noddies in a flock while feeding on bait fish.

water, up the center line, and when it has passed the limit of the school flies up and around either to the left or right and rejoins the body of the flock, and repeats the sequence which may take as little as 15 seconds or as long as 2 minutes. The flock can move in any direction while holding this pattern; it can go forward, to the left, to the right, backwards, along diagonals, and so on. The center line faces into the wind.

But a flock is also chaotic, which causes interference between birds. Prey are uniformly distributed in the area where the flock is feeding, at least until the predatory fish begin feeding on them. Noddies concentrate at places where the prey break the surface. Thus one sees a blanket of noddies covering these places with many others flying or hovering above unable to get access to the prey. Noddies frequently threaten each other while feeding, especially when avoiding midair collisions. This description makes it seem that noddies caught above the blanket are severely disadvantaged. Indeed they are for that short period, but the unpredictability of where the bait fish will break the surface may ensure that all birds get access to them at some time. As Feare (1981) pointed out, feeding rates of seabirds in flocks are probably limited by space available for feeding. This appears to be the case in flocks of noddies.

Within 30 seconds of the bait fish school diving, the flock begins to disperse. One of the outcomes of this is that there are likely to be some birds above the place where predatory fish bring bait fish to the surface. Noddies reaggregate over the bait fish. The flock lasts provided the fish are brought to the surface frequently. If prey are unavailable for more than a few minutes, noddies start to leave the area. Some noddies rest on the surface in rafts when bait fish are unavailable (Serventy, Serventy, and Warham 1971). They rise off the water and rejoin the flock when the bait fish are available. At times predatory fish split into two groups, each working a section of the school. Hence the bait fish and the flock are each split. The two resulting flocks have lives of their own but may be reunited if the bait fish reform one school. The composition of a flock is frequently changing because birds often leave and others join it.

The noddy's feeding techniques are well suited to catching many prey quickly during the short time that they are available. For example, noddies often plunge to the surface feet first and catch one to three fish in their bills while remaining on the surface for a few seconds. Sometimes noddies hover but in all cases they remain close to the surface.

The frequency with which each seabird species forages with noddies probably reflects their respective abilities to adjust to foraging with them (table 3.4). Black-naped terns modify their foraging methods by plunging to the surface from less than 1 meter. So black-naped terns can operate from below the blanket of birds but they do not seem to be able to catch as many prey as noddies do. The number of black-naped terns decreases with the size of the noddy flock (r = −0.555, df 101, P < 0.001). This relationship may indicate that the availability of prey to the non-noddy species is too low to warrant foraging with large flocks. The interference between foraging birds may cause a decrease in the availability of prey.

Crested, lesser crested, roseate, and bridled terns do not adjust their foraging methods as much as black-naped terns do. Hence these species continue to operate from above the blanket without much success. Indeed these species tend to forage toward the edge of a flock where there are fewer noddies and perhaps a lower density of prey, both of which may be more suited to their foraging methods. Crested terns approach the school from above the noddies, then slowly descend between the birds and plunge when a path is cleared.

Both black-naped and roseate terns form flocks, at times, over schools being pursued by Spanish mackerel *(Scomberomorus commersoni)* and shark mackerel *(Grammetorcynus bicarinatus)*. Mackeral do not concentrate the prey as much as tuna do and this, perhaps, allows the terns a higher capture rate than when they are hunting prey concentrated by tuna because noddies do not join large numbers. In a sample of fourteen of these flocks, the mean number of noddies in them was eighteen (s.d. = 8).

Crested, lesser crested, and bridled terns may join feeding frenzies close to the reef, but the density of birds is much less than in most noddy flocks, and the birds fly several meters above the water. These frenzies occur when predatory fish chase bait fish into the shallow waters of the surf.

Prey Size and Type
The preferred prey size of a predator that eats its prey whole is largely affected by the width of its gape (Hulsman 1981). Kislalioglu and Gibson (1976) reported that the optimal prey size of the fifteen spined stickleback *(Spinachia spinachia)* was about half the width of the largest prey it could swallow. Preferred prey size is that which the predator eats in preference to other sizes, whereas optimal prey size is that which yields the highest net gain in energy per unit of handling time (Kislalioglu and Gibson 1976). Preferred and optimal prey sizes, in practice, seem to be the same (see Kislalioglu

and Gibson 1976; Swennen and Duiven 1977). This principle also applies to seabirds (Swennen and Duiven 1977; Hulsman 1981). Since relaxed width of gape is probably proportional to its maximum width one can theoretically determine the preferred prey size of each individual of each species. Therefore one could generate a frequency distribution of preferred prey sizes of each prey species for each predator species.

Unfortunately I do not have sufficient information to do this for the terns at One Tree Reef, but we can get a general picture from the mean widths of their gapes and lengths of their bills. The width of the gape affects the size of prey that a predator can swallow, whereas the length of the bill affects the size of prey that it can catch (see Hulsman 1981). Birds with long bills will usually catch smaller prey than ones with shorter bills because the force exerted at the bill tip decreases as the length increases unless the strength of the adductor muscles increases to compensate (see Ashmole 1968).

The relationship between bill morphology and prey size of the terns that co-occur at One Tree Reef is as follows. Noddies, black-naped, and roseate terns are about the same body size and therefore could have the same amount of adductor muscle and hence strength to close their bills. But these three species have very different bill morphology (table 3.5). Noddies have longer and more slender bills than roseate terns, which have slightly longer bills than black-naped terns. Noddies have smaller gapes than roseate and black-naped terns. On the basis of this information, I surmise that noddies will have smaller preferred prey than will roseate terns, which will

Table 3.5. Mean measurements (± 1sd) for six tern species.

| SPECIES | CHARACTER | | |
	Bill length (mm)	Width of gape (mm)	Weight (g)
Crested	62.2 ± 3.0	18.5 ± 1.8	320
Lesser crested	51.3 ± 1.3	16.4 ± 0.9	240
Black-naped	34.6 ± 1.6	12.5 ± 1.0	100
Roseate	37.4 ± 1.8	12.3 ± 0.8	115
Bridled	41.9 ± 2.0	13.2 ± 1.2	131
Black noddy	44.8 ± 1.8	12.1 ± 0.8	114

have smaller preferred prey than black-naped terns. Bridled terns, although larger than noddies and roseate and black-naped terns, are smaller than lesser crested terns, which are smaller than crested terns in body size, bill length, and width of gape (table 3.5). If the amount and strength of adductor muscle increases with body size then despite the longer bills, crested terns would eat larger prey than would lesser crested terns, which would eat larger prey than would bridled terns. This general trend in preferred prey sizes is consistent with the distribution of prey sizes eaten by these six tern species. This is clearly shown in the prey category called Silver species (table 3.6).

Black-naped terns do not seem to fit the predicted pattern. Their prey was on the average smaller than that caught by roseate terns and noddies. Perhaps the roseate terns and noddies do fit the predicted pattern, although roseate terns and noddies caught a wider range of prey sizes than back-naped terns caught at the time these data were collected (table 3.6). At other times of the breeding season, both roseate terns and noddies caught significantly smaller prey than black-naped terns did (t = 6.94 and 7.52, respectively, df = 1,060 and 1,083, P < 0.001; Hulsman 1977). This may indicate that their preferred prey sizes were not common enough where these seabirds foraged and they therefore supplemented their diets with prey from less-preferred size classes.

The type of prey that each species eats is a function of its nutrient requirements, bill morphology, and the prey's availability where the tern species forages. For example, crested terns ate a greater range of sizes and variety of prey than other tern species (table 3.6). Some prey types were simply too large or wide for the smaller terns to swallow. In addition, crested terns can dive deeper than the others and therefore may have a larger available prey population from which to choose.

Time of Breeding

According to Perrins' model, species breed when they can meet their nutrient and energy requirements. Therefore if the available prey are closer to the preferred prey size of one tern species than of another, all other things being equal, the former will breed earlier than the latter. This is because the former could build up its nutrient and energy reserves at a faster rate than the other species could at that time. This scenario in part matches what occurs at One Tree Reef where the breeding seasons of some species are staggered (figure 3.9).

Table 3.6. Percentages of prey types in the diet of six tern species at One Tree Island.

SPECIES OF TERN	LENGTH OF PREY (CM)									
	> 0–2	> 2–4	> 4–6	> 6–8	> 8–10	> 10–12	> 12–14	> 14–16	> 16–18	> 18–20
Silver Species (surface)										
Crested			10.7	13.9	12.1	5.2	5.1	3.7	0.8	0.5
Lesser crested		46.3	9.8	14.1	7.6	8.7	5.5	1.1		
Black-naped			41.0	12.6						
Roseate	1.1	6.9	57.5	32.2	2.3					
Bridled			4.3	4.3		4.3				
Black noddy	2.4	29.0	38.6	30.0						
Carangidae (surface)										
Crested				0.2	1.2	0.7		0.1		
Exocoetidae (surface)										
Crested				0.3	0.42	0.5	0.36	2.4	0.85	0.5
Bridled				2.2	2.2					
Balistidae (surface)										
Crested			0.24	1.2	1.4	1.5	1.0			
Bridled		10.9	43.4	19.5	6.5					
Tetraodontidae (surface)										
Crested			0.7	3.6	4.7	0.6	0.12			

Pomacentridae (midwater and bottom)							
Crested			0.3	0.5	0.54	0.5	0.5
Labridae (midwater and bottom)							
Crested	0.06	0.12	3.6	1.2	0.8	0.3	0.12
Scaridae (midwater and bottom)							
Crested			0.06		0.3	0.42	0.12
Blenniidae (bottom)							
Crested				1.2	0.9	0.3	0.24
Lesser crested				16.3	6.5	12.0	2.2
Bridled							2.2
Gobiidae (bottom)							
Crested	1.8	0.24	0.9	1.5	1.8		

Lesser crested terns breed in spring in years that their main prey species, *Pranesus capricornensis*, is common (Hulsman 1979). Black-naped terns sometimes breed in spring when atherinids and anchovies (4 to 6 cm long) are common. Roseate terns also breed in spring, usually about two weeks after black-naped terns begin. Roseate terns are slightly larger (table 3.5), probably have smaller preferred prey, and travel farther from their colonies when foraging than black-naped terns do (figure 3.6). Therefore it takes roseate terns slightly longer than black-naped terns to build up their nutrient and energy reserves to the level to start breeding when prey are closer to the size preferred by black-naped than roseate terns.

Noddies start breeding in early summer, that is, well after black-naped and roseate terns, even though noddies are about the same size as roseate terns. The later breeding by noddies may result from their having smaller preferred prey and traveling farther from their colonies in search of food. If the prey are larger than the preferred prey size of noddies, they may not be able to utilize them as efficiently and thus take longer to build up their reserves to meet their breeding needs. Another important factor in the time at which

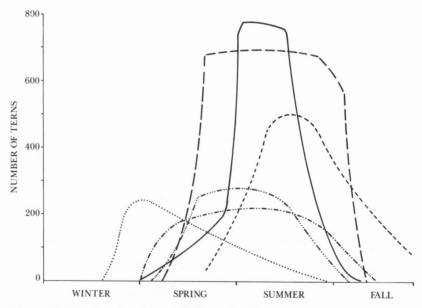

Figure 3.9. Number of each species of tern breeding at One Tree Island during a year; crested tern (———); lesser crested tern (.....); black-naped tern —··—); roseate tern —····—; bridled tern (— —); black noddy (- -).

noddies breed is the activity of predatory fish. In years in which the bait fish are not common and hence predatory fish are not common in the area, fewer noddies breed and they breed later than when bait and predatory fishes are common.

Bridled and crested terns begin breeding in late spring. Although bridled terns are smaller then crested terns they travel much farther from their colonies in search of food. This may account for why bridled terns do not start breeding before the larger crested terns.

The significance of staggered breeding is that periods of maximum food demand (i.e., when feeding young) do not coincide for more than a few species at any one time. The fact that this sequence is not always followed is consistent with the hypothesis that the onset of breeding is related to the availability of the birds' prey. This relationship has been demonstrated in the kestrel (Falco tinnunculus) (Dijkstra et al. 1982). Most species have left the area by the beginning of the fall (figure 3.9). Their departure could be caused by a number of factors: 1) a diminished food supply, 2) an increased demand for food by a larger population (adults plus fledglings), 3) a decrease in available food, or 4) increased competition for access to food.

I will deal with each of these points in order. The birds' departure is not likely to be caused by a diminished food supply because the recruitment of reef fish is still close to its peak in the beginning of the fall (see Russell, Anderson, and Talbot 1977). Even though there are more birds in the area than there were during nesting, the food supply should be at or near its peak and sufficient to meet the needs of the birds (see figures 3.3 and 3.9).

A more likely explanation for their departure is that there is a decrease in the available food supply, even though the total food supply has increased. Availability may decrease because of the poor weather conditions in the area during late summer. Strong winds, which make sea surface conditions difficult for seabirds to catch prey, occur more often during late summer than at any other time. Also, cyclones are more likely to affect the weather during late summer than at any other time (Lourensz, in Langham 1986). Some species are more affected by poor weather than others. For example, cyclonic weather reduced the fledging success of bridled terns to 77 percent, whereas it reduced that of crested terns to 24 percent in 1980 at One Tree Island. Ordinarily both of these species have high fledging success (> 80 percent; Hulsman and Langham 1985).

Seabirds disperse northward from the Capricornia Region, according to recoveries of banded birds (Hulsman, unpublished data). They may not escape the poor weather but their density is lower than near their colonies and, therefore, there is less potential competition for food. This hypothesis, however, remains to be tested and at this stage we do not know whether competitive exclusion or some other reason causes their departure.

■
Conclusions

Perrins' model not only provides a mechanism for timing of breeding but also for structuring seabird communities. Competition for food is likely to affect birds in Regions 1 and 2 during the breeding season when they are concentrated in a small area and food is not superabundant. In contrast, competition for food is likely to affect birds in Regions 3 and 4 during the nonbreeding season when food is scarce.

This could well resolve the apparent discrepancy between Lack's hypothesis and Ashmole's hypothesis as to when competition for food should affect bird populations. Lack (1966) argued that competition for food during the nonbreeding season controlled seabird populations, whereas Ashmole (1971) argued that in the tropics if competition for food controlled seabird populations it had to be during the breeding season. Let us consider the four hypothetical patterns of resource abundance. The pattern of resource abundance in Region 1 is similar to that in the tropics (see Ashmole 1971). The pattern in Region 2 is like that in the subtropics (see Kinsey 1972). The patterns in Regions 3 and 4 are like those in temperate and polar regions respectively. If this is indeed true, it would mean that Ashmole's hypothesis applies in tropical and subtropical localities and Lack's hypothesis applies in temperate and polar localities.

Sympatric species of terns co-occur via an interplay of morphological differences and resulting abilities: the form of bill, and body size (Ashmole 1968); the color of plumage (Phillips 1962); maneuverability in flight (Ashmole and Ashmole 1967; D. C. Duffy, personal communication); the ability to hunt in the presence or absence of predatory fish (Hulsman 1978) or dispersed or concentrated prey (Erwin 1977).

Prey type and size were the most important variables in distinguishing guilds of terns (Hulsman 1987). The niche of a species is dynamic and it could well be this dynamism that enables closely related species to co-occur (Hulsman 1987). The size and type of prey in a bird's diet are a function of the bird's morphology, foraging methods, foraging zones, selection of prey, interactions with other birds and predatory fish, and behavior and growth of prey. Decreased availability of food caused by poor weather conditions during late summer could increase competition for access to food and lead to adults and fledglings leaving the colony. Although the birds still may be affected by poor weather conditions elsewhere, their lower density means less competition for access to food.

■

Summary

A theoretical framework of the structure of seabird communities is presented. The framework is constructed from general principles derived from Perrin's model, which describes the nature of the association between food supply and time of breeding. It is this framework which provides the context in which the structure of seabird communities is discussed.

Four hypothetical patterns of food abundance are used in conjunction with Perrin's model to predict the timing of molt and breeding of seabirds as well as the role of competition for food in affecting community structure. For example, in regions where the food supply varies little during the year but remains above the level at which breeding can be sustained, molt and breeding must be separated. Moreover the time required to complete molt may affect the actual time of breeding. Also competition for food is likely to affect birds during the breeding season when they are concentrated in a small area and food is not abundant. In contrast, in regions where food is superabundant for a small portion of a year, e.g. four months, molt and breeding may co-occur and many species breed at the same time. Competition for food is likely to affect birds during the nonbreeding season when food is scarce.

Seabird communities of the Great Barrier Reef are used to answer two questions: what factors affect the number of species in a locale? and what factors affect the number of species belonging to each feeding and breeding category (inshore, offshore, and pelagic) in a locale?

Sympatric species of seabirds co-occur via an interplay of morphological differences and resulting abilities. Prey type and size were the most important variables distinguishing guilds of terns. These characteristics of their diets are a function of the birds' morphology, foraging methods, foraging zones, selection of prey, interactions with other kinds and predatory fish, behavior, and growth of prey.

Acknowledgments

I thank David Duffy, Don Kinsey, and Jan Veen for sowing the seeds for the concepts that are presented here. I am most grateful to Hanneke, Jan, Jort, and Thor Veen for their hospitality and a quiet corner while I wrote this synthesis. Also I thank Joanna Burger and Carl Safina for their useful comments about the manuscript. Figure 3.2 was drawn by Aub Chandica; figures 3.3, 3.5, and 3.8 were drawn by Lyn Coccettii; figure 3.7 was drawn by Karen Papecek. My thanks to B. Coates for the use of his unpublished observations on tuna feeding on bait fish. I also thank Shena Melkus for typing the various drafts of this manuscript.

References

Ainley, D. G. and R. J. Boekelheide. 1983. An ecological comparison of oceanic seabird communities of the South Pacific Ocean. *Studies in Avian Biology* 8:2–23.

Ashmole, N. P. 1968. Body size, prey size, and ecological segregation in five sympatric tropical terns (Aves: Laridae). *Syst. Zool.* 17:292–304.

Ashmole, N. P. 1971. Seabird ecology and the marine environment. In D. S. Farner and J. R. King, eds., *Avian Biology*, 1:223–286. New York: Academic Press.

Ashmole, N. P. and M. J. Ashmole. 1967. Comparative feeding ecology of seabirds of a tropical oceanic island. *Peabody Mus. Nat. Hist. Bull.* 24:1–132.

Burger, J. 1981. A model for the evolution of mixed-species colonies of Ciconiformes. *Quart. Rev Biol* 56:143–167.

Dale, P. E. R., K. Hulsman, B. R. Jahnke, and M. Dale. 1984. Vegetation and nesting preferences of Black Noddies at Masthead Island, Great Barrier Reef. Part 1: Patterns at the macro-scale. *Aust. J. Ecol.* 9:343–351.

Diamond, A. W. 1978. Population size and feeding strategies in tropical seabirds. *Amer. Nat.* 112:215–223.

Dijkstra, C., L. Vuursteen, S. Daan, and D. Masman. 1982. Clutch size and laying date in the Kestrel *Falco tinnunculus:* Effect of supplementary food. *Ibis* 124:210–213.

Erwin, R. M. 1977. Feeding and breeding adaptations to different food regimes in three seabirds: The Common Tern *Sterna hirundo,* Royal Tern, *Sterna maxima,* and Black Skimmer, *Rynchops niger. Ecology* 58:389–397.

Feare, C. J. 1981. Breeding schedules and feeding strategies of Seychelles seabirds. *Ostrich* 52:179–185.

Hulsman, K. 1977. Feeding and breeding biology of six sympatric species of tern (Laridae) at One Tree Island, Great Barrier Reef. Ph.D. dissertation, University of Queensland, Australia.

Hulsman, K. 1978. Reactions of fish to hunting methods of terns: a means of segregation. *Proc. Colonial Waterbird Group* (New York 1978), 2:105–109.

Hulsman, K. 1979. One Tree Island, Queensland. *Corella* 3:38–40.

Hulsman, K. 1980. Feeding and breeding strategies of tropical terns. *Proc. 17th Intern. Ornith. Congr.* (Berlin 1978), p. 984–988.

Hulsman, K. 1981. Width of gape as a determinant of size of prey eaten by terns. *Emu* 81:29–32.

Hulsman, K. 1987. Resource partitioning among sympatric species of tern. *Ardea.* 75:255–262.

Hulsman, K. In preparation. A model of foraging methods of terns in relation to depths to which they submerge.

Hulsman, K. and N. P. Langham. 1985. The breeding biology of the Bridled Tern *Sterna anaethetus. Emu* 85:240–249.

Hulsman, K., N. P. Langham, and D. Bluhdorn. In press. Factors affecting the diet of Crested Terns *Sterna bergii. Aust. Wildl. Res.*

Kikkawa, J. 1976. The birds of the Great Barrier Reef. In O. A. Jones and R. Endean, eds., *Biology and Geology of Coral Reefs,* vol. 3: *Biology 2,* pp. 279–341. New York: Academic Press.

Kinsey, D. W. 1972. Preliminary observations on community metabolism and primary productivity of pseudo-atoll reef at One Tree Island, Great Barrier Reef. *Proc. Symp. Coral and Coral Reefs* 1969:13–32.

Kislalioglu, M. and R. M. Gibson. 1976. Prey "handling time" and its importance in food selection by the 15-spined stickleback *Spinachia spinachia* (L.). *J. Exp. Mar. Biol. Ecol.* 25:151–158.

Kruuk, H. 1964. Predators and anti-predator behaviour of the Black-headed Gull *(Larus ridibundus* L.). *Behaviour Suppl.* 11:1–129.

Lack, D. L. 1966. *Population Studies in Birds.* Oxford: Claredon.

Lack, D. L. 1968. *Ecological Adaptations for Breeding in Birds.* London: Methuen.

Langham, N. 1986. The effect of cyclone "Simon" on terns nesting on One Tree Island, Great Barrier Reef, Australia. *Emu* 86:53–57.

Lemmetyinen, R. 1971. Nest defence behaviour of Common and Arctic Terns and its effects on the success achieved by predators. *Ornis Fenn.* 48:13–24.

Nisbet, I. C. T. 1975. Selective effects of predation in a tern colony. *Condor* 77:221–226.

Phillips, G. C. 1962. Survival value of the white colouration of gulls and other seabirds. D. Phil. thesis, Oxford University.

Perrins, C. M. 1970. The timing of bird's breeding seasons. *Ibis* 112:242–255.

Russell, B. C., G. R. V. Anderson, and F. H. Talbot. 1977. Seasonality and breeding of One Tree Reef fishes. *Aust. J. Mar. Freshwater Res.* 28:521–528.

Searley, S. G. 1973. Interspecific feeding assemblages of marine birds off British Columbia. *Auk* 90:796–802.

Serventy, D. L., V. Serventy, and J. Warham. 1971. *The Handbook of Australian Sea-Birds.* Sydney: Reed.

Schodde, R., B. Glover, F. C. Kinsky, S. Marchant, A. R. McGill, and S. A. Parker. 1978. Recommended English names for Australian birds. *Emu Supplement;* 245–314.

Simmons, K. E. L. 1972. Some adaptive features of seabird plumage types. *Br. Birds* 65:465–479, 510–521.

Swennen, C. and R. Duiven. 1977. Size of food objects of three fish-eating seabird species: *Uria aalge, Alca torda* and *Fratercula arctica* (Aves, Alcidae). *Netherlands J. Sea Res.* 11:92–98.

Ward, P. and A. Zahavi. 1973. The importance of certain assemblages of birds as "information centres" for food finding. *Ibis* 115:517–534.

■ Competition and Predation

4

Ecological Dynamics Among Prey Fish, Bluefish, and Foraging Common Terns in an Atlantic Coastal System

Carl Safina · *National Audubon Society*
Islip, New York
Joanna Burger · *Department of Biological Sciences*
Rutgers University

The Importance of Food Distribution Patterns

The spatial and temporal distribution of food is a fundamental factor affecting the evolution of a species. It is one of the most important determinants of where an animal fits on the continuum between solitary territoriality and group living, and in the development of coloniality among birds in particular (Horn 1968; Cody 1974; Wilson 1975). For seabirds, food distribution patterns are vitally important in an array of life history characteristics; in shaping coloniality (Ward and Zahavi 1973; Erwin 1978); in the phenology of reproduction (Perrins 1970; Veen 1977; Gochfeld 1980); in determining reproductive success (e.g., Nisbet 1973; Courtney and Blokpoel 1980); and ultimately in limiting seabird numbers (Ashmole 1963; Furness 1982). Yet few studies describing direct observation of population dynamics and interactions among seabirds, prey fishes, and predatory fishes on the foraging grounds have been published (Erwin 1983a), because data on fish have been largely lacking. Harrison, Hida, and Seki (1984) remark that many fundamental questions regarding seabird biology cannot be answered until prey availability is better understood. Theoretical relationships between colony size, colony location, species composition, timing of breeding, and available food (Ashmole 1963; Lack 1968; Nelson 1970; Ward and Zahavi 1973; Burger 1981) cannot be fully tested until food resources can be examined. The implications of such untested theoretical relationships apply throughout avian taxa. Surprisingly, even for noncolonial and terrestrial birds, current views on the role of food in determining breeding and social organization remain largely untested (Schluter 1984).

In this paper we review past studies of food availability to piscivorous birds and present new data on the dynamics among fish as they relate to common tern foraging. We develop models of interactions of the major components of the system in the hope of stimulating studies of interactive dynamics rather than the more usual autecological approach to investigating the lives of individual seabird species.

Past Studies

Studies of the relationship between seabirds and their prey have been carried out largely by researchers who are primarily ornithologists (thus they usually have a more bird-oriented, rather than fish-oriented, perspective) and have usually been based on prey brought to chicks at nests, stomach contents, or regurgitated pellets (e.g., Harris 1965; Pearson 1968; Hunt and Hunt 1976; Baltz, Morejohn, and Antrim 1979; Courtney and Blokpoel 1980). The relationship of prey provisioning rates has been studied with respect to mate selection, nesting synchrony, egg quality, clutch size, chick growth, and colony size and composition (e.g., Perrins 1970; Nisbet 1973, 1977, 1978; Nisbet and Cohen 1975; Parsons 1976; Veen 1977; Erwin 1978; Gochfeld 1980; Ryder 1980; Burger 1981; Powell 1983). The observation of prey brought to colonies cannot shed light on the pre- or postbreeding availability of food. Further uncertainties in relying on the rate of prey provisioning to assess food availability are introduced by the fact that parents may adjust the prey they bring back based on the age of chicks (Nisbet 1981; Miller and Confer 1982) or on nutrient requirements. Thus, studies of food brought to chicks cannot usually give a clear idea of what the parents are catching.

Recently, studies have been published on various aspects of fishing behavior by birds, such as social attraction, intra- and interspecific resource partitioning, age-related differences in fishing success, and rhythms of feeding activity (e.g., Dunn 1972; Hopkins and Wiley 1972; Ward and Zahavi 1973; Sealy 1973; Buckley and Buckley 1974; Kushlan 1976, 1977; Pratt 1980; Hoffman, Heinemann, and Wiens 1981; Bayer 1981; Atwood and Kelly 1984; Gochfeld and Burger 1982; Burger and Gochfeld 1983; Porter and Sealy 1982). These studies of fishing behavior have largely addressed themselves to those aspects of food acquisition that occur above water. Kushlan (1978) pointed out that, although studies of food habits of water birds remain useful, more critical questions involve the way prey is selected from the range of potential prey available and the role of such factors as prey density, species, and size in determining prey selection.

Though it is widely recognized that the availability of food on the fishing grounds is a major factor influencing distribution, reproductive success, and colony site selection, it has been the subject of few studies in the past because of difficulties in sampling prey (Ogden and Nesbit 1979; Courtney and Blokpoel 1980; Buckley and Buckley 1980). Several studies have sought ways around this problem by incorporating independent fisheries data (Erwin 1977; Briggs et al. 1981; Anderson, Gress, and Mais 1982; Miller and Confer 1982; Duffy 1983; Schaffner 1986). Other studies have been carried out in shallow waters, where conventional nets and traps were operable (Kushlan 1976, 1979; Ogden, Kushlan, and Tilmant 1976; Black and Harris 1983; Hafner and Britton 1983). In deep oceanic water, studies of the relationship between food distribution and bird ecology began with avian species that forage on plankton, because certain types can be sampled with relative ease (Brown 1980; but see Ashmole 1971). These studies found that prey availability varied, causing responses in foraging behavior by seabirds, and that availability of prey was sometimes a product of prey abundance and other factors that concentrated prey and facilitated prey capture. These were significant advances in understanding predator behavior, prey selection, and predator-prey population dynamics, but the sampling methods they used could not be applied to piscivorous birds in deep-water situations, where fishes and squids are highly mobile and patchily distributed and are adept at avoiding sampling nets. Sealy (1973) examined the prey available to mixed seabird flocks off British Columbia by towing plankton nets "with limited success." Small fish upon which the birds preyed were seldom obtained in these samples. The ability of small fish to avoid nets has been demonstrated (Pitcher and Wyche 1983).

In the past, lack of useful data on fish dynamics repeatedly hampered understanding of bird/fish trophic relationships. Kushlan (1979), in a paper on white ibis *(Eudocimus albus)*, said that "much remains to be learned about the relationship between availability of prey in natural systems and its consumption by predators." Buckley and Buckley (1980), discussing colony site selection by terns, noted a lack of fisheries data and said that while food availability plays a major role in habitat selection, it is a subject that remains to be unraveled. In a review of postfledging parental care in seabirds, Burger (1980) pointed out that data on food availability were generally lacking. Erwin (1983a), in his recent synthesis on feeding behavior and ecology of colonial water birds, reiterated that "almost no effort has been made at assessing changes in fish communities." Salt and Willard (1971) observed a decline in overall success over the season and speculated about prey population changes, but they lacked data to test their ideas about the latter. Murphy et al. (1984) documented changes in the food taken by breeding glaucous-winged gulls *(Larus glaucescens)* during the breeding season and related this to reproductive parameters in the gulls. They thought that the reproductive success of the birds in their study was profoundly food limited, but they lacked data on population changes in food species between a year when they inferred high food availability and a year when they inferred low availability; thus their conclusions, although quite reasonable, could not be confirmed.

These studies indicated a need to gather data on food dynamics concurrent with observations of the birds themselves, if food fluctuations and the mechanisms through which birds were influenced by, and responded to, food variability were to be understood. Our work was designed to help fill this need (Safina and Burger 1985; Safina and Burger, ms a–c). Several years ago we recognized the potential of using sonar to study prey populations of fish-eating seabirds in deep water. Sonar has been used in quantitatively assessing pelagic stocks of small fish, especially anchovies *(Engraulis mordax)*, for a number of years (e.g., Hewett, Smith, and Brown 1976; Koslow 1981). Briggs et al. (1981) and Anderson, Gross, and Mais (1982) were among the first to show that this technique could be applied to the study of seabirds, when they included sonar surveys of anchovies, done independently for purposes of fisheries management, in analyses of brown pelican *(Pelecanus occidentalis)* distribution. Since 1982 we have been using sonar to quantify population dynamics of prey fish eaten by common terns, factors facilitating prey availability, and bird-fish competition between common terns and predatory fish. The major advantage of our method is that we can gather data on fish population dynamics directly. It allows construction of a quantitative index of fish school abundance and distribution in situ without the biases and disrupting effects that nets or other gear have on fish behavior. Consequently, sonar data represent a relatively accurate profile of the distribution, depth, and density of fish in the water column. Ours is the only study we are aware of that is designed specifically to quantify fish population changes as they relate to seabirds, assessing prey population dynamics and availability by working directly among feeding birds. The methodology has shown itself to be an effective tool in understanding intra- and interannual food supply dynamics, prey fish behavior, competition and commensalism between terns and predatory fish, and tern foraging behavior (Safina and Burger, in press).

The Question of Optimality

Questions revolving around whether foraging behavior allows an animal to make optimal use of resources have billowed into one of the most important fields within ecological theory (Engen and Stenseth 1984). Optimal foraging theory is an attempt to find out if there are any general rules about what animals feed on, where they search, and what decisions they use (Krebs, Stephens, and Sutherland 1983). Optimization models have been qualitatively successful, because of their modest aims, but have been less successful quantitatively (Krebs, Stephens, and Sutherland 1983). The usefulness of optimization models in field situations has been limited by the enormous complexity of the environment within which animals function (Zach and Smith 1981). It may appear tautologous to speak of optimizing in the context of natural selection (Cody 1974). In fact, however, many animals do not, and perhaps cannot, always behave in ways that allow the optimization (most efficient use) of some currency function such as time, energy, or nutrients. Because natural selection will presumably provide a best mix of solutions to an array of problems rather than a best solution to any one problem (Janetos and Cole 1981), and because animals are confronted with constraints and tradeoffs (Myers 1983), the world seems full of suboptimal or imperfect animals, at least when perfection is judged by the degree to which a facet of an animal's life matches the predictions of researchers.

Studies demonstrating energetic inefficiency, what may be termed "suboptimal foraging," are building a substantial literature (e.g., Janetos and Cole 1981; Munger 1984; Erwin 1985), and have resulted in growing dissatisfaction with optimality theory in general

(Heinrich 1983; Engen and Stenseth 1984; Glasser 1984). Engen and Stenseth suggest "something is wrong in current foraging theory,' and Glasser asserts that "most empiricists are still testing a theory that should have been modified or replaced long ago." When observations are interpreted in terms of, and forced to fit, the conventional paradigm despite strains in the data, we risk allowing the theory to dictate our view of the world rather than using it as a tool to increase our understanding of nature (Glasser 1984). Perhaps most insightfully, Heinrich (1983) suggested that rapid progress could be made by examining the mechanisms that act to enhance foraging returns, without worrying about whether or not the behavior is optimal. Janzen (1986) has admonished that nature is more interesting than are our predictions of what nature should be like.

Sih (1982) distinguishes between "optimization," which maximizes fitness, and "energy maximization," a strategy for maximizing net energy intake, which he says is what foraging theory now generally is. We would add that, given the complexity of real habitats (considering the effects of imperfect knowledge and the array of other things that an animal must do to survive), the best advantage that can accrue to an animal attempting to maximize is simply the enhancement of energy gain (or whatever currency) over what would be encountered if the search pattern was random.

With all this in mind, in looking at tern foraging, we have generally avoided the temptation to be fashionable and to couch our questions and interpretations in terms of optimality. Had we done so, we would have found, like many studies, general qualitative agreement, with some exceptions. In our approach to understanding the relationship between terns and fish, we have sought to shed light on these questions: What are the dynamics of food in the ambient environment? What are the conditions affecting food availability to terns? Do terns respond to their ambient environment in ways that result in enhanced encounter rates with capturable prey?

We examined a nearshore oceanic system on the continental shelf off the northeastern United States. Common terns arrive in midspring and begin breeding almost immediately. Reproductive activities usually occur through July, when the terns begin to disperse. While in these waters, terns prey on a variety of small fish species whose numbers and relative abundances differ markedly within and among years. The dominant predatory fish in the system is the bluefish *(Pomatomus saltatrix)*, which arrives several weeks after the terns do and has profound effects on the numbers and behavior of prey fish. It was our aim to explore and gain an understanding of the interactive dynamics among prey fish, bluefish, and common terns.

■

Methods

We conducted our field studies from May through August during 1982–85 in the New York Bight (northwest Atlantic Ocean) near Fire Island Inlet, New York (40°N, 73°W). Fire Island Inlet is approximately 2 kilometers east of a large tern colony at Cedar Beach (figure 4.1). The Cedar Beach colony is located on a sand and shingle barrier beach where the principal vegetation is beach grass *(Ammophila breviligulata)*, seaside goldenrod *(Solidago sempervirens)*, and sea rocket *(Cakile edentulata)*. The common tern population at this colony has grown from 2,500 pairs in 1980 to 6,000 pairs in 1986. This appears to be the largest colony of common terns in the world at present (M. Gochfeld, personal communication). Approximately 80 pairs of roseate terns *(Sterna dougalii)*, 200 pairs of black skimmers *(Rynchops niger)*, and several pairs of willets *(Catoptrophorus semipalmatus)*, killdeer *(Charadrius vociferus)*, and piping plovers *(Charadrius melodus)* breed among the common terns. Within 5 kilometers of Fire Island Inlet are several small salt marsh colonies totalling approximately 1,000 pairs of common terns.

Data were collected aboard a 5.5-meter boat equipped with loran C and a Raytheon DC 200 paper-recording echo sounder. The sounder

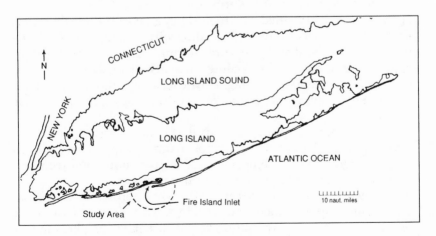

Figure 4.1. The study area.

was calibrated over clear water where fish could be seen (figure 4.2). Prey fish, predatory fish, and terns usually ignored the boat during transects and may have been habituated to boats in this area, which is frequented by fishermen in small craft.

We ran two types of transects, control and flock. Control transects were run along the same 0.5-kilometer square route near the inlet mouth several times a day, at intervals of two hours or more, whether or not birds were present. Flock transects were run through foraging flocks of terns and as far past the flock as the flock was wide. This allowed the water column under feeding terns to be compared with an adjacent area of equal length and identical sea conditions but without terns. Flock transects were run anywhere in the ocean where birds fed within an approximately 10-kilometer radius of the inlet.

For each transect, date, time, tidal phase, wind speed, water clarity (measured with a Secci disk), surface water temperature, air temperature, sea surface choppiness, and an estimate of the number of foraging terns in the area were recorded. Any common terns or predatory fish seen within 20 meters of the boat were noted manually on the sounder's paper. Sea surface choppiness was classified as either: 0 (slick surface), 1 (light ruffled surface), 2 (normal chop), or 3 (breaking chop, or whitecaps). Prey fish were usually identified as they were carried from the water by terns, by directly observing them in the water, or by examining the stomachs of bluefish and northern weakfish *(Cynoscion regalis)*. For predatory fish, size, species, and an index of feeding intensity were determined by trolling multiple-tube lures on wire during each transect. Activities of predatory fish were classified as either: 0 (no activity), 1 (feeding deep, i.e., with no surface activity), 2 (feeding with sporadic surface activity), or 3 (prolonged surface feeding during which many fish chased prey at the surface). The feeding intensity of predatory fish was classified as either 0-no feeding, 1-little feeding, 2-moderate feeding, or 3-intense feeding, depending on the frequency with which fish struck lures.

An index of tern feeding activity level and capture success was obtained immediately after each sonar transect by observing adult common terns through binoculars for 5–10 minutes and recording successful and unsuccessful fish capture attempts.

Field work was begun in early May when terns first arrived on the breeding grounds and terminated in August after most birds had left the area. Date of termination varied, depending on phenology of departure. Observations were usually begun at dawn (approximately

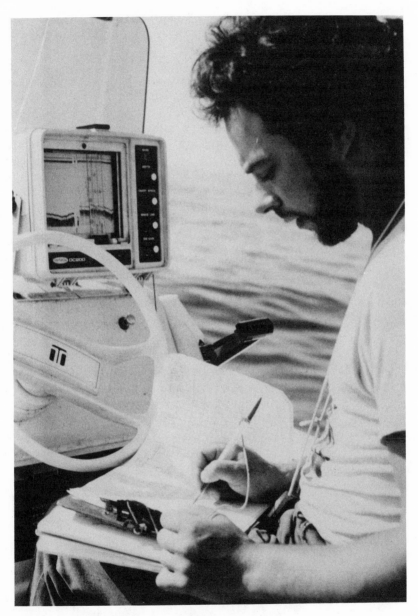

Figure 4.2. Safina recording data with field equipment.

5:00 A.M.) and terminated around noon. Data were generally collected from three to five days per week, depending on weather conditions. Nearly six hundred sonar transects were run over the four years of field work.

For each transect, fish densities were quantified by overlaying the echo sounder paper with a transparent 7-millimeter square grid and estimating the percent coverage of prey and/or predatory fishes in each grid square. Predatory and prey fishes were differentiated by their echo marks. Echoes of predatory fish tend to form discrete spikes, while prey fish schools appeared as dark, irregularly shaped masses (figure 4.3); see also Safina and Burger 1985). This method allowed a variety of useful indices of fish density, abundance, school size, and fish depth to be calculated. Density was defined as the mean percent coverage of echo marks per grid square for the entire transect. Density was calculated as the sum of percent coverages in all grid squares with echo marks, divided by the total number of grid squares with and without echo marks. Thus, density could be equal in very short and very long transects. Abundance, defined as an index of biomass, was calculated as density multiplied by length of transect. Because school dimensions are related to the number of fish (Pitcher and Partridge 1979), an index of school size, based on the relative area covered by fish groups, was calculated as the mean percent coverage of echo marks in grid squares with marks. Fish depth was defined as the mean depth of fish in the echo profile. Sonar records were divided vertically into discrete 1.5-meter vertical depth segments. To calculate fish depth, fish density was multiplied by depth for each depth segment, these values were summed,

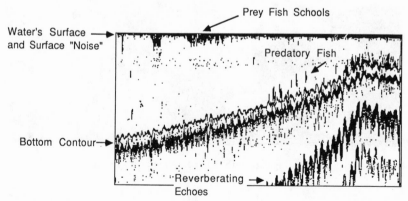

Figure 4.3. Portion of a sonar transect.

and this sum was divided by the sum of percent coverage of echo marks in each grid square. This yielded a depth-weighted mean.

Data were analyzed using SAS computer programs. Due to their schooling nature, fish were usually either absent or present in substantial numbers, resulting in a bimodal distribution of data that was resistant to normalizing transformations. For this reason, nonparametric analyses were usually performed.

■

Results

The Relationships Between Fish and Physical Phenomena

Variability in physical aspects of the ocean can affect fish abundance and location. Mean water clarity and water temperature differed significantly among years ($\chi^2 = 54.06$, df = 3, $p < .0001$; $\chi^2 = 5.35$, df = 1, $p < .02$, respectively). Clarity averaged highest in 1983, followed in order by 1984, 1982, and 1985. Clarity was substantially lower in 1985 than in the other three years, possibly as the result of the mysterious "brown tide," a bloom of chrysophyte algae (yet to be fully identified) which first appeared in Long Island estuaries in 1985 (Cosper et al. 1987). Water temperature was measured for only two years, and was warmer in 1984 than in 1985. Water temperature was inversely correlated with prey fish density ($tau = -.19$, $p < .01$), abundance ($tau = -.21$, $p < .004$), and school size ($tau = -.18$, $p < .02$), though the degree to which this was related to date and the effects of predators is difficult to disentangle. Prey and predator depth were correlated with clarity ($tau = .09$, n = 231, $p < .04$; $tau = .22$, n = 105, $p < .002$, respectively) (figure 4.4), indicating that fish prefer to remain away from the surface when visibility is high.

We measured a number of physical variables concurrent with our sonar transects. In flock transects, where water depth varied, prey abundance increased with water depth ($tau = .19$, n = 274, $p < .00001$) (figure 4.5). Predator abundance did not increase with water depth, nor did prey or predator density or school size, but predator size did ($tau = .50$, n = 125, $p < .0001$). Both mean prey depth and mean predator depth were correlated with water depth ($tau = .45$, n = 250, $p < .0001$; $tau = .68$, n = 117, $p < .0001$, respectively). This is evidence of a preference by fish to remain away from the surface. Prey density in the upper 3, 4.5, 6, and 7.5 meters were inveresly related to water depth, but the difference ceased to be significant after 6 meters from the surface (table 4.1). This suggests that prey prefer to stay deeper only up to a point, and that although they avoid the surface they also tend to avoid the bottom (figure 4.6).

We then compared the effects of other physical variables on fish in flock versus control transects. Tide did not correlate with any fish variable for control transects, but in flock transects several variables responded to tidal changes. Prey fish density and abun-

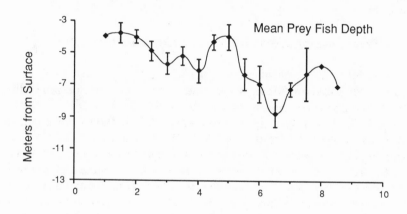

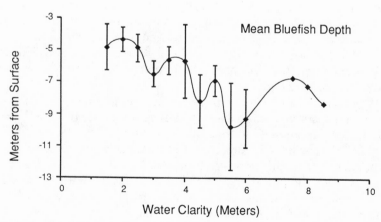

Figure 4.4. Prey fish and predatory fish depth as a function of water clarity (means ± SE).

Table 4.1. Correlations of mean prey fish depth with water depth in the upper levels of the water column (n = 276).

	tau	$p <$
Upper 3.0 m	−.19	.00001
Upper 4.5 m	−.16	.0002
Upper 6.0 m	−.10	.02
Upper 7.5 m	−.04	.4

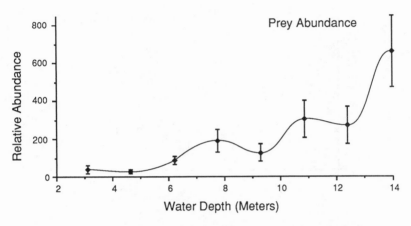

Figure 4.5. Changes in mean prey fish abundance ($\pm$ SE) with water depth.

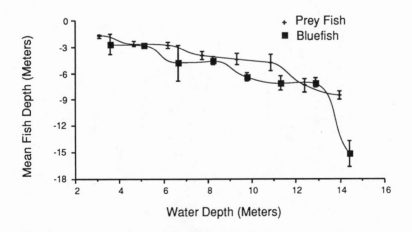

Figure 4.6. Changes in mean fish depth ($\pm$ SE) as a function of water depth.

dance varied among tidal stages (high, ebb, low, flood) (for density, Kruskal-Wallis test, $X^2 = 11.23$, df = 3, $p < .01$; for abundance, $\chi^2 = 8.34$, df = 3, $p < .04$, respectively) and was highest on ebbing tide (figure 4.7). Prey fish density and abundance in the upper 3 meters of the water column was likewise highest on ebbing tide (Kruskal-Wallis test, $\chi^2 = 10.06$, df = 3, $p < .02$; $\chi^2 = 9.15$, df = 3, $p < .03$, respectively). Mean prey depth and school size did not change with tide. Bluefish were most likely to appear around low

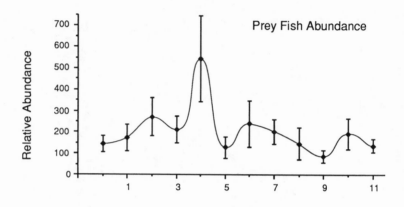

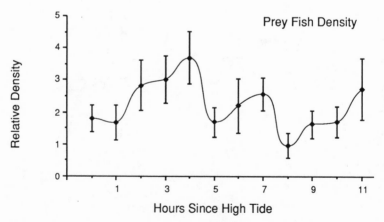

Figure 4.7. Changes in prey fish density and abundance over the tidal cycle (means ± SE).

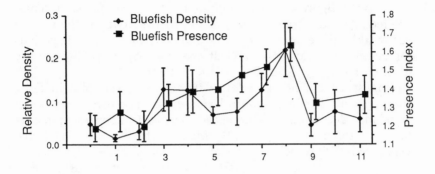

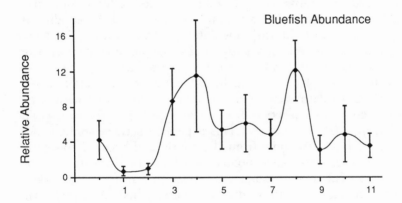

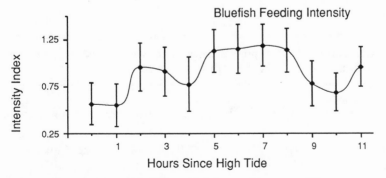

Figure 4.8. Responses of bluefish to tidal changes (means ± s SE).

tide (χ^2 = 8.65, df = 3, $p < .03$). Bluefish density and abundance were greatest during ebb and low incoming water (for density, χ^2 = 7.59, df = 3, $p < .05$; for abundance χ^2 = 6.53, df = 3, $p < .09$, respectively; figure 4.8). Tide appeared not to affect predator depth. The feeding intensity of predators and their activity at the surface was weakly higher around the low and early incoming tide periods (for feeding intensity, χ^2 = 6.83, df = 3, $p < .08$; for feeding activity at the surface, χ^2 = 5.60, df = 3, $p < .1$; figure 4.7).

Wind speed did not correlate with any fish variables in flock transects, but it did in control transects. Prey fish density and abundance decreased as wind speed increased in control transects ($tau = -.15$, n = 261, $p < .001$; $tau = -.14$, n = 261, $p < .003$, respectively). Prey school size also decreased as wind increased ($tau = -.15$, n = 233, $p < .002$). These three factors suggest that fish density and abundance close to shore decreases with increasing wind speed, but terns are still able to locate at least some fish schools to forage over, at least up to the moderate wind conditions (to approximately 24 km/hr) in which we worked. Mean prey depth did not change significantly with wind speed, but prey density near the surface decreased, and the effect tended to be strongest close to the surface (table 4.2). Thus, at higher winds, fish tended to avoid the surface. Predator density and abundance correlated with wind speed in control transects ($tau = .11$, n = 261, $p < .03$; $tau = .11$, n = 261, $p < .03$, respectively). This is perhaps because strong winds were most likely to come from the south, and because southerly winds often seem to "push" predatory fish toward shore (toward the control transect area), through their effect on warmer surface water masses. Predator activity, feeding intensity, school size, and mean depth were not significantly correlated with wind speed.

Cloud cover did not affect fish in flock transects, nor did it affect

---

Table 4.2. Correlations of prey density with wind speed at different distances from water surface (Kendall's *tau*) n = 261.

Meters from surface:

	tau	$p <$
3.0	−.11	.02
4.5	−.10	.04
6.0	−.09	.07
7.5	−.09	.05

prey fish in control transects, but it did have an effect on predatory fish in control transects. Predators were more likely to be present in controls as cloud cover increased (tau = .16, n = 260, p < .002). Predator density and abundance increased with increasing cloud cover in control transects (tau = .16, n = 260, p < .002 for both comparisons). Predator feeding intensity and feeding activity near the surface also increased with cloud cover (tau = .13, n = 261, p < .01; tau = .10, n = 262, p < .06, respectively), indicating that bluefish were more likely to move closer to shore into shallower water, where we ran controls, when the sky was overcast. Predator depth (distance from the surface) and school size were unaffected by cloud cover.

Sea surface choppiness had an effect on fish. In flock transects, where prey had more vertical latitude than in control transects, prey depth increased with increasing choppiness (tau = .10, n = 242, p < .05). No other prey variable in either transect type responded to changing choppiness, but most bluefish variables did (table 4.3). Of the variables that responded to choppiness, bluefish presence, abundance, and density were more depressed by choppiness in flock transects than in controls, while their feeding was more depressed in controls. This is probably best explained by the relationship between bluefish feeding and the tracking of bluefish by terns; nonfeeding bluefish are not likely to appear in flock transects because terns do not congregate over them. That bluefish fed less when seas were more choppy is best shown in controls, because these transects were run in the same location regardless of the activity of fish or terns. When bluefish were not feeding vigorously due to choppiness, they were often nonetheless present, as is indicated by the fact that bluefish presence, abundance, and density did

---

Table 4.3. Correlations between sea surface choppiness and predatory fish variables in Flock and Control transects (Kendall's tau).

	Control n ≈ 269		Flock n ≈ 260	
	tau	p <	tau	p <
Bluefish presence	−.11	.07	−.15	.01
Abundance	−.10	.09	−.14	.01
Density	−.10	.09	−.14	.01
Feeding intensity	−.18	.002	−.08	.15
Surface feeding activity	−.20	.0006	−.11	.04

not respond significantly to choppiness in control transects. Mean depth and school size of bluefish were also unaffected by choppiness in either transect type.

Water temperature was correlated with fish variables. Prey density was inversely related to water temperature in both control and flock transects ($tau = -.19$, n = 88, $p < .01$; $tau = -.19$, n = 108, $p < .006$, respectively). So was prey abundance ($tau = -.21$, n = 88, $p < .004$; $tau = -.22$, n = 108, $p < .0008$, respectively). In flock transects, prey depth was inversely related to water temperature ($tau = -.13$, n = 102, $p < .07$), but there was no relation between water temperature and fish depth in control transects, perhaps because the moderately shallow water and restricted range in depth provided less opportunity for depth changes. Prey school size was inversely related to water temperature in both control and flock transects ($tau = -.18$, n = 88, $p < .02$; $tau = -.18$, n = 101, $p < .009$, respectively). Predators showed no correlation with water temperature in control transects, but they did in flock transects, where every variable except school size correlated with water temperature (table 4.4).

It is difficult to disentangle the extent to which predator-prey interactions and date caused the observed responses to water temperature rather than changing temperature per se over the course of the season. To see if water temperature warming facilitates the arrival of fish, mean air temperatures in March through May (Brookhaven National Laboratory, unpublished) were compared with temporal patterns of prey fish maxima and predatory fish arrival. Results are equivocal (table 4.5). Control and flock transects were examined with attention to week of prey abundance maxima. Flock

---

Table 4.4. Correlations between sea surface temperature and predatory fish variables in Flock transects (Kendall's *tau*).

	Flock Transects, n ≈ 110	
	tau	p <
Bluefish presence	.16	.06
Abundance	.16	.04
Density	.17	.03
Feeding intensity	.23	.002
Surface feeding activity	.24	.002

transects were also examined with attention to the week of first arrival of detectable numbers of bluefish. Control transects were not examined with respect to bluefish because no bluefish appeared there in 1984. The rank of March and April temperatures in 1983–85 corresponded with the rank of week of peak prey numbers in control transects and the arrival of bluefish in flock transects. May temperature ranks corresponded with ranks for week of prey maxima in flock transects. Thus it appears that fish arrive earlier in years when weather is warmer than usual.

The Relationship of Tern Foraging Activities to Fish

Foraging flock size and the total number of terns foraging in the area are indices of the response of terns in the local population to changes in food availability. Throughout the breeding season, a relatively static number of locally nesting adults appear and disappear from the foraging grounds in numbers that vary dramatically within and among days. We sought to provide information on the relationship between fish activity and the patterns of tern presence and distribution on the foraging grounds.

Flock Size and Fish

Tern flock size was strongly correlated with prey abundance (table 4.6; figure 4.9) and prey density, especially density within 3 meters of the surface. When we compared flock size with prey

Table 4.5. Rank-order relationships of March–April mean air temperatures to prey fish abundance maxima and predatory fish arrival.

	1983	1984	1985
Prey fish peak week (rank)			
Flock	4 (earliest)	5 (mid)	6 (latest)
Control	4 (mid)	5 (latest)	2 (earliest)
Bluefish arrival week (rank)	4 (mid)	7 (latest)	3 (earliest)
Mean air temps in °C			
March	5.0 (mid)	0.7 (lowest)	4.7 (highest)
April	8.7 (mid)	8.4 (lowest)	10.0 (highest)
May	12.4 (lowest)	12.8 (mid)	14.8 (highest)

NOTE: First week of May is week 1, etc.

density in the upper 4.5, 6, and 7.5 meters of the water column, we continued to find highly significant correlations. Flock size also correlated strongly with prey school size. There was no general relationship between flock size and mean prey depth. When terns were foraging over bluefish schools, flock size was larger when bluefish were shallower. There was no general relationship between tern flock size and predatory fish density, abundance, or school size. Flock sizes were much larger, however, when bluefish were feeding intensely and were much in evidence at the surface (figure 4.10).

We produced a number of fish variable interactions involving different prey and predator variables to see if we could find correlations that were higher than those produced by the variables we measured directly. We suspected that prey depth and abundance were both important in determining availability to terns, and that the presence of predatory fish would increase availability through the tendency of predatory fish to cause a decrease in prey depth. Prey abundance divided by mean prey depth improved upon our correlations between flock sizes and single variables, as did multiplying the prey abundance by predatory fish presence (predatory fish presence was assigned a value of 2, while predatory fish absence was assigned a value of 1) (table 4.6).

Table 4.6. Correlations between flock size, total number of birds in the study area, and fish (Kendall's Correlation).

	Flock Size			Total Birds		
	tau	n	p <	tau	n	p <
Prey abundance	.20	266	.00001	.18	215	.0001
Density	.17	266	.0001	.17	215	.0003
School size	.15	239	.0007	.15	193	.003
Density in the upper 3.0 m	.17	265	.0001	.11	216	.02
Density in the upper 4.5 m	.16	265	.0003	.10	216	.03
Density in the upper 6.0 m	.15	265	.0004	.09	216	.06
Density in the upper 7.5 m	.15	265	.0005	.10	216	.03
Abundance/prey fish depth	.23	237	.00001	.20	192	.0001
Predator mean depth	−.20	113	.003	−.10	89	.2
Prey abund. × predator presence	.21	265	.00001	.19	214	.00001

Fish Distribution and the Numbers of Terns
Feeding in the Study Area

Presumably, certain features of fish distribution and behavior might differentially affect the numbers and grouping of terns within the study area. The total number of birds present might correspond differently to fish variables than would their flock size distribution, resulting in different numbers of flocks for the same total number of birds or vice versa. This was not the case, however. Flock sizes and the total number of birds feeding in the study area correlated with fish variables in a similar manner, except that flock size responded to the feeding activities and depth of predatory fish, while total numbers of foraging birds did not. As with flock size, the mean number of birds in the study area was correlated with prey abundance, prey density, and mean prey school size (table 4.6). Also, as with flock size, the overall number of birds in the study area was correlated with prey density in the upper 3, 4.5, 6, and 7.5 meters of the water column. As with flock size, there was no general relationship between numbers of birds in the study area and mean prey depth. There was also no correlation between total bird numbers and bluefish density, abundance, school size, feeding activities, or mean depth. These last two variables (predator feeding and predator depth) are the only ones for which flock size was correlated but overall number of birds in the study area was not. This suggests that the importance of bluefish to the foraging ecology of terns is through the feeding activities of bluefish; terns in the area strongly

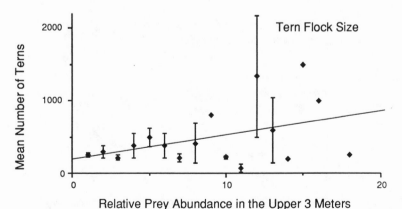

Figure 4.9. Mean (± SE) flock sizes of terns in response to changes in prey abundance in the upper 3 meters of the water column.

aggregate to sites where bluefish are feeding, resulting in larger flocks even though the total number of terns in the area is not significantly affected. As with flock size, prey abundance divided by mean prey depth proved to be a better predictor of number of birds in the study area than were single variables, as was multiplying the prey abundance by two if predatory fish were present (table 4.6).

Flock Density and Fish

Flock density (the number of terns divided by transect length) showed some different relationships to fish than did flock size and

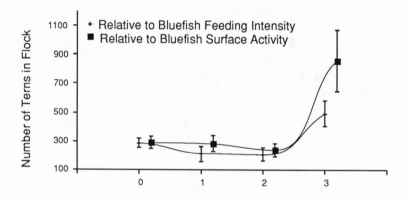

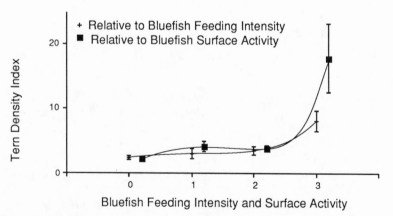

Bluefish Feeding Intensity and Surface Activity

Figure 4.10. Responses of terns to bluefish feeding intensity and surface activity. Bluefish feeding intensity and surface activity are measured in relative terms, scored 0 to 3, as described in the section on methods (means ± s SE).

numbers of feeding terns. Density was inversely correlated with prey abundance ($tau = -.13$, n = 269, $p < .002$), indicating that birds preferred to spread out when prey was abundant. Flock density was correlated with prey density in the upper 3 meters of the water column ($tau = .08$, n = 271, $p < .04$), but not with prey density at 4.5, 6, or 7 meters from the surface, or overall prey density, probably

---

Table 4.7. Prey brought to tern nests at Cedar Beach.

Common taken prey in all or most years:

American sand lance	*Ammodytes americanus*
Bay anchovy	*Anchoa mitchilli*
Bluefish	*Pomatomus saltatrix*
Butterfish	*Peprilus triacanthus*

Commonly taken in some years; regular but uncommon in others:

Atlantic mackerel	*Scomber scombrus*
Long-finned squid	*Loligo pealei*
Round herring	*Etrumeus teres*
Blueback herring	*Alosa aestivalis*
Atlantic herring	*Clupea harengus*
Scup	*Stenotomus chrysops*

Uncommon but regularly taken prey:

Mummichog	*Fundulus heteroclitus*
Striped killifish	*Fundulus majalis*
Common pipefish	*Syngnathus fuscus*
Atlantic menhaden	*Brevoortis tyrannus*
Winter flounder	*Pseudoplueronectes americanus*

Rarely taken prey:

Common shore shrimp	*Palaemonetes vulgaris*
Mole crab	*Emerita talpoida*
Moth	unidentified
Atlantic silverside	*Menidia menidia*
Atlantic saury	*Scomberesox saurus*
Glasseye snapper	*Priacanthus cruentatus*
Goosefish	*Lophius americanus*
Atlantic moonfish	*Vomer setapinnis*
Windowpane	*Scophthalmus aquosus*
Lined seahorse	*Hippocampus erectus*
Striped croaker	*Bairdiella sanctaeluciae*
Halfbeak	*Hyporhamphus unifasciatus*
Silver hake	*Merluccius bilinearis*
Common shiner	*Notropis cornutus* (a freshwater sp.)
Goldfish	*Carassius auratus* (a freshwater sp.)
Sunfish	*Lepomis* spp. (a freshwater sp.)

because within a flock birds track and congregate over the densest groups of fish they can see. Prey depth had an important effect on tern density, with density increasing as prey averaged closer to the surface (tau = .22, n = 244, p < .00001).

Tern density correlated strongly with almost every measure of predatory fish abundance and behavior: bluefish abundance (tau = .23, n = 268, p < .00001), density (tau = .27, n = 268, p < .00001), closeness of bluefish to the surface (tau = .16, n = 114, p < .01), amount of activity at the surface (tau = .28, n = 266, p < .00001), and bluefish feeding intensity (tau = .27, n = 265, p < .00001) (figure 4.9). Tern density also strongly responded to the mere presence or absence of predatory fish (tau = .27, n = 268, p < .00001). Density of terns in feeding flocks was not, however, related to bluefish school size.

Tern Diving and Prey Fish

Common terns are generalist feeders and capture a wide variety of prey (table 4.7). Total fish-oriented behaviors (completed dives, aborted dives, and hovers combined) related significantly to prey density (tau = .09, n = 252, p < .04), and less so to prey abundance (tau = .08, n = 252, p < .06). Considered individually, the number of successful, unsuccessful, and aborted dives correlated with prey density (tau = .09, n = 255, p < .04; tau = .10, n = 255, p < .02; tau = .12, n = 255, p < .005, respectively). This was the only statistically significant correlation between successful dives and single prey fish variables (though interactions of some variables produced other significant correlations with tern fishing success rate; see the section on variable interactions). Weak, statistically insignificant correspondence between success and prey density at discrete distances from the surface grew even weaker as prey densities at increasing depths were considered. Dividing prey density by mean prey depth was an improved way of predicting tern diving success (tau = .11, n = 220, p < .02). Looking at correlations between prey abundance and successful (tau = .07, n = 255, p < .1), unsuccessful (tau = .11, n = 255, p < .01), and aborted dives in turn (tau = .15, n = 255, p < .0005), the relationship grows progressively stronger. One might suspect that the reason that numbers of unsuccessful and aborted dives increase with prey abundance is simply that terns dive more when there is more prey, but the fact that only a very weak correlation was found between prey density and successful dives, while stronger correlations existed between prey density and unsuccessful and aborted dives, may suggest that terns are less "careful" when food is abundant.

The number of dives initiated (successful, unsuccessful, and aborted combined) and the number of dives completed were correlated with prey density, prey abundance, and the product of prey density and abundance (table 4.8). Terns hovered less as prey went deeper (tau = .16, n = 222, p < .001), probably because hovering mostly occurs after prey have been sighted, and they also hovered less as prey abundance increased (tau = .11, n = 249, p < .01), probably because prey abundance and mean prey depth were positively correlated. With increasing depth of prey, the number of unsuccessful dives tended weakly but not significantly to increase (tau = .08, n = 221, p < .1). There were no significant correlations between prey school size and diving activities of terns.

In addition to looking at numbers of dives by terns, we also examined the percentages of total fish-oriented behaviors that were successful dives, unsuccessful dives, aborted dives, or hovers. There was no statistically significant relationship between any prey variable tested and the percent of completed dives that were successful. The relationships between prey density and abundance and the percent of dives that were successful were weak (tau = .08, n = 247, p < .06; tau = .07, n = 247, p < .12, respectively). Because the percent of dives that were successful was generally stable, the number of successful and unsuccessful dives correlated strongly with the number of dives completed (tau = .67, n = 260, p < .0001; tau = 84, n = 261, p < .0001). The percent of aborted dives was negatively but not significantly correlated with prey density in the upper 3 meters of the water column and was positively correlated with prey

Table 4.8. Correlations among prey density, abundance, depth, and tern diving frequency.

	Dives Initiated			Dives Completed		
	tau	n	*p* <	*tau*	n	*p* <
Prey density	.12	243	.01	.10	243	.02
Prey abundance	.14	243	.002	.10	243	.02
Density × abundance	.13	243	.002	.11	243	.02
Density/prey depth	.10	215	.02	.09	215	.05
Abundance/prey depth	.14	215	.002	.11	215	.02

density in the upper 7.5 meters (tau = .09, n = 247, p < .04). The percent of fish-oriented actions that were hovers was inversely related to prey density, prey abundance, and prey depth (tau = −.13, n = 241, p < .003; tau = −.19, n = 241, p < .00001; tau = −.22, n = 241, p < .00001, respectively), but not to school size.

The number of agonistic interactions (threatening vocalizations and chases) among terns was correlated with factors that indicate poor prey availability near the surface, but there was no correlation with overall prey density, overall abundance, or school size per se. Agonistic encounters correlated with increasing prey depth (tau = .17, n = 178 p < .003), and they were negatively correlated with the density of prey in the upper 3 and 4.5 meters of the water column (tau = −.12, n = 195, p < .03 for both comparisons). Agonistic interactions were also correlated with sea surface choppiness (tau = .14, n = 208, p < .03), which may indicate poor visibility for foraging terns and which causes prey to go deeper. The rate of agonistic encounters was inversely related to the percent of dives that were successful (tau = −.12, n = 183, p < .04) and was correlated to the number of aborted dives (tau = .10, n = 206, p < .05).

Tern Diving and Predatory Fish

We examined tern diving in relation to the abundance, density, school size, mean depth, and feeding activities of bluefish. The number of successful and unsuccessful dives were not significantly correlated with these bluefish variables. However, the number of all dives taken together (successful, unsuccessful, and aborted) was inversely related to predator presence and density (tau = −.18, n = 242, p < .0006; tau = −.13, n = 242, p < .008). The number

Table 4.9. Correlations between number of dives aborted by terns and predatory fish abundance, density, feeding intensity, and feeding activity near the surface (Kendall's *tau*).

	Number of Aborted Dives		
	tau	n =	p <
Predatory fish presence	−.19	248	.0003
abundance	−.15	248	.002
density	−.14	248	.005
feeding intensity	−.14	265	.003
surface activity	−.15	267	.002

of aborted dives was inversely related to predator presence, abundance, density, feeding intensity, and feeding activity at the surface (table 4.9), but not with mean predator depth or school size. The number of completed dives (successful + unsuccessful) was greater when predators were present (Kruskal-Wallis test, $\chi^2 = 4.53$, df = 1, $p < .03$) but was not correlated to predator density. Terns dove less when bluefish were present. However, they tended to complete more of the dives they initiated (Kruskal-Wallis test, $\chi^2 = 2.47$, df = 1, $p < .1$) and have a higher percent capture rate per dive (Kruskal-Wallis test, $\chi^2 = 3.56$, df = 1, $p < .06$) (figure 4.11). It is this relationship of tern diving to bluefish presence that partially mitigated the depressive effect of bluefish on prey numbers (figure 4.12) and the relationship of dwindling prey numbers to prey availability over the course of the season. Thus the effect on terns of the seasonal decline in prey numbers, which we believe is ultimately attributable to the predatory activities of bluefish, is retarded by the bluefish themselves in their unusual role of "commensal competitors" (figures 13–15). Overall, terns' mean prey capture rate was the same whether bluefish were present or absent ($\chi^2 = 0.01$, df = 1, $p < .9$; figures 4.11 and 4.16). Averaged over prey density categories, tern fishing success reached its highest level in the absence of bluefish, whereas averaged over weeks, their success peaked in the presence of bluefish (figure 4.13). This emphasizes the fact that there were no consistent differences in terns' prey capture rates based on the presence or absence of bluefish. The maximum success

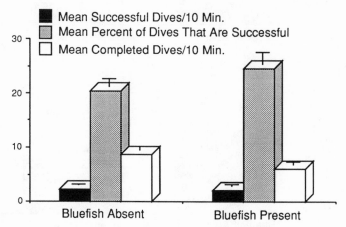

Figure 4.11. Tern success rate, percent success, and diving rate in areas where bluefish were absent compared to areas where they were present (means ± SE).

rate of terns in the absence of bluefish was 1.5 fish captures/minute; in the presence of bluefish the terns' maximum success rate was similar: 1.2/minute.

Terns did much hovering when bluefish were feeding in the vicinity (figure 4.17). The impression we got was that terns hovered while they waited for bluefish they could see to attack prey and drive it to the surface. Hovering at these times was often close to

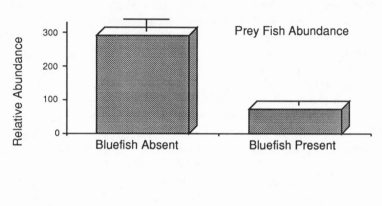

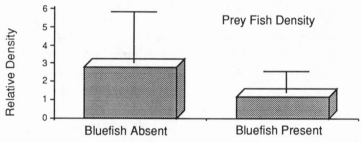

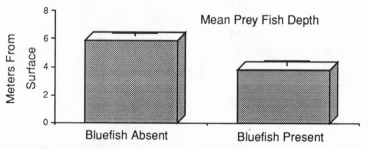

Figure 4.12. Prey fish responses to the presence of bluefish (means ±s SE).

the water's surface, and terns often moved forward while hovering. Hovering increased markedly with bluefish abundance and density (table 4.10) and with bluefish feeding intensity and activity at the surface. Terns hovered less as bluefish went deeper. Hovering was not related to bluefish school size. The percent of dives that were aborted was not related directly to most bluefish variables. The percent of fish-oriented behaviors that were hovers were related to the same bluefish variables as were number of hovers (table 4.10).

Terns foraging over actively feeding predatory fish can be distin-

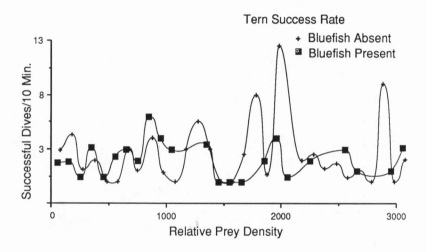

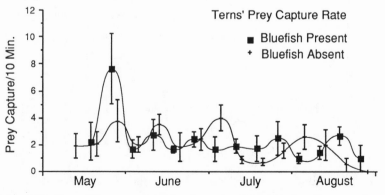

Figure 4.13. Changes in mean fish abundance (± SE) under foraging terns during summer.

Table 4.10. Correlations between tern hovering, percent of fish-oriented behaviors that were hovers, and Bluefish variables (Kendall's *tau*).

	Number of Hovers			Percent Hovers		
	tau	n =	*p* <	*tau*	n =	*p* <
Predator abundance	.18	248	.0002	.22	240	.00001
density	.20	248	.00001	.24	240	.00001
feeding intensity	.24	264	.00001	.28	256	.00001
surface activity	.22	266	.00001	.27	258	.00001

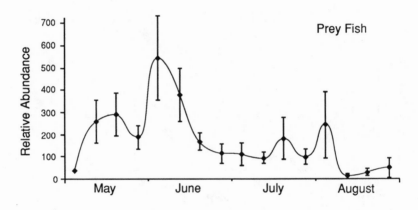

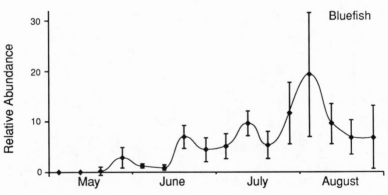

Figure 4.14. Prey abundance in flock transects with versus without bluefish during summer (means ± s SE).

guished at great distance from those feeding in areas where there are few or no predators by the amount of hovering they do and their dense distribution in small irregularly spaced groups whose members focus much attention on small patches of water. The result is often a group of relatively funnel-shaped flocks of terns hovering excitedly, the funnel often moving tornadolike. Much vocal threatening can be heard as the terns jostle for position, occasionally colliding. The rate of agonistic interactions among terns was correlated to the amount of hovering ($tau = .16$, $n = 206$, $p < .005$). When bluefish surface in a relatively large school, the tern flock

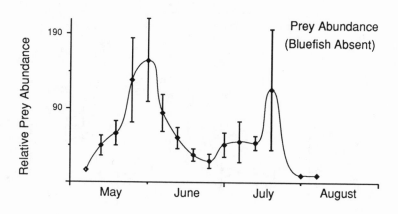

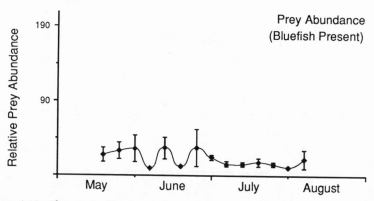

Figure 4.15. Changes in dive rate and prey capture success of terns during summer (means ± s SE).

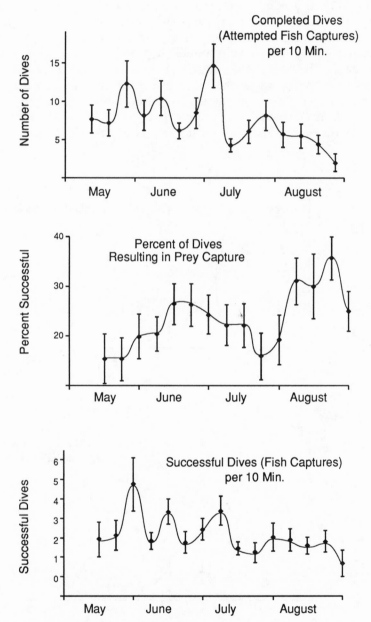

Figure 4.16. Success rates of terns at different prey densities and different times of the season in the presence and absence of bluefish (means ± s SE).

resembles a dense cloud. Very high dive rates can occur in these situations, with the peculiar visual effect that it is "raining" terns. This effect is caused by the fact that terns present a more visible surface during dives, when they are perpendicular to the observer's line of sight, than during their ascent back into the flock, when they are primarily parallel to the observer's line of sight. Such clouds seldom last for more than 30 seconds. If the predatory fish are appearing at the surface sporadically, terns spread out to search and then converge rapidly, often from hundreds of meters, when one or more fish appears or when one tern stops to hover or dive. Under these conditions only the nearest terns have a high chance of

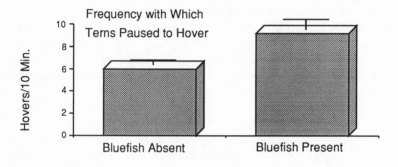

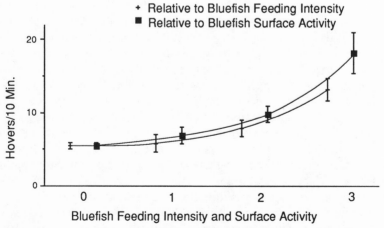

Figure 4.17. Tern hovering in response to bluefish presence, feeding intensity, and feeding activity at the surface (means ± s SE).

catching a fish before the bluefish disappear, although terns that approach, even if not successful, may be closer to the place that the bluefish will surface next. This pattern is the rule in deeper waters. Common terns rarely fished in water more than 15 meters deep unless predatory fish were feeding in the area.

The presence of bluefish changes the relationship between the availability of prey to terns and prey density and abundance. Correlative relationships between prey fish and tern diving behavior change depending on whether bluefish are present or absent. When bluefish were present under feeding terns, the success of terns was related more to prey abundance (tau = .14, n = 167, $p < .01$) than density (tau = .10, n = 167 $p < .08$). When bluefish were absent, the relationship was reversed, with success related to density (tau = .19, n = 89, $p < .01$) but not abundance (tau = .12, n = 89, $p < .13$).

Prey depth did not increase with time of day, nor did prey density, but prey abundance tended to be reduced over time during the morning if bluefish were not present (tau = $-.10$, n = 158, $p < .06$). If bluefish were present, prey abundance was not reduced over the course of the morning (tau = $-.04$, n = 112, $p < .5$).

Prey/Predator Variable Interactions and Tern Foraging

Our observations on the water led us to suspect that certain variables interacted in concert in making prey available to terns. Consequently, we tested a number of interactions among fish variables and compared them to foraging behaviors of terns (table 4.11). The table shows single-variable correlations as well, for comparison with the interaction variables. The numbers of dives correlated with the variable interactions in a number of cases. In general, however, the interaction variables did not improve greatly on the correlative relationships between foraging activities and the single fish variables. For number of successful dives, dividing density of prey by prey depth improved upon the correlation between number of successful dives and prey density. For the number of dives that terns aborted, predator feeding intensity and predator presence tended to magnify the relationships between prey and rate of aborted dives. One thing that becomes evident in the table is that the more removed from a successful dive the behavior is, the more it responds to some variable or variables. That is to say, aborted dives were more responsive to changes in fish variables than were unsuccessful dives, which were more responsive than successful dives. The number of successful dives thus seems "buffered" against the variability in most fish variables.

The Relationship of Tern Foraging Activities to Physical Variables

Physical variables are likely to play a role in the availability of food to terns, either because they change the nature of the air/water interface, or because they affect fish behavior. In general, factors favoring low light penetration into the water seem to favor tern foraging. Birds oriented to fish (dove and hovered) less as water

Table 4.11. Interactions among prey abundance, prey density, prey depth, predators, and tern diving.

| | Tern Dives | | | | | | | | |
| | Successful | | | Unsuccessful | | | Aborted | | |
	tau	n	*p* <	*tau*	n	*p* <	*tau*	n	*p* <
Prey density	.09	255	.04	.10	255	.01	.12	255	.005
Prey abundance	.07	255	.1	.11	255	.01	.15	255	.0005
Predator feeding intensity	−.03	274	.6	−.05	274	.3	−.15	274	.002
Prey depth	.02	226	.7	.09	226	.06	.09	226	.06
Predator presence	.01	274	.9	.11	274	.04	.19	274	.0003
Prey density/ prey depth	.11	220	.02	.07	220	.2	.11	220	.02
(Prey density/ prey depth) × predator presence	.11	219	.03	.05	219	.3	.07	219	.1
Prey density × predator presence	.10	247	.03	.07	247	.02	.09	247	.04
Prey abundance/ prey depth	.10	220	.05	.10	220	.04	.17	220	.0003
Prey abundance × predator presence	.08	247	.08	.09	247	.05	.13	247	.03
Prey abundance × predator feeding intensity	−.05	246	.3	−.12	246	.02	−.19	246	.0001
Prey density × predator feeding intensity	−.04	246	.4	−.11	246	.02	−.18	246	.0001

clarity increased ($tau = -.09$, n = 248, $p < .04$). They made a smaller number of completed dives ($tau = -.09$, n = 246, $p < .04$) and they aborted a higher percentage of initiated dives as clarity increased ($tau = .10$, n = 238, $p < .02$). The number of successful dives was negatively but not significantly related to clarity. Water clarity was negatively related to tern density ($tau = .12$, n = 249, $p < .009$) but not to flock size or numbers of terns in the area. Tern density also increased with increasing cloud cover ($tau = .11$, n = 266, $p < .01$), but flock size and numbers of terns in the area did not. Cloudiness and sea surface conditions were unrelated to tern foraging success. Sea surface choppiness, at least in the calm to moderate conditions at which we usually worked, was unrelated to tern numbers or density.

As the sun rose higher, flock size and density and the number of birds in the area decreased ($tau = -.29$, n = 285, $p < .0001$; $tau = -.16$, n = 265, $p < .0001$; $tau = -.31$, n = 235, $p < .0001$, respectively; figure 4.18). So did the number of successful dives and total fish-oriented behavior ($tau = -.11$, n = 261, $p < .01$; $tau = -.09$, n = 261, $p < .02$, respectively).

Wind speed was positively related to tern density ($tau = .14$, n = 260, $p < .003$), but it was unrelated to numbers of foraging terns. It was not significantly related to the number of successful dives (table 4.12), although it was negatively correlated to the percent of completed dives that were successful ($tau = -.12$, n = 222,

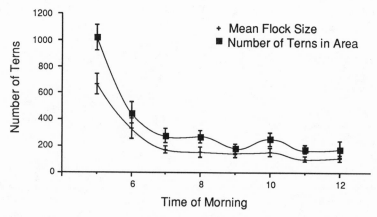

Figure 4.18. Numbers of foraging birds as a function of time of morning (means ± SE).

$p < .02$). Interactions between prey density, prey abundance, and wind speed suggest that wind may inhibit the terns' ability to capture prey. Prey density and prey abundance divided by wind speed resulted in closer correlations with tern success rates than either of the prey variables alone. However, dividing by wind speed depressed the correlation between prey density and unsuccessful dive rates, and did not change the correlation between prey abundance and the rate of unsuccessful dives. These interaction variables improved the correlation between the percent of dives that were successful and prey density and abundance, although these relationships were not statistically significant (table 4.12). Sea surface choppiness, which is often a function of wind speed, was not significantly related to diving or flock variables.

Most tern activity was greater during ebb tide, but the relationships were statistically weak (figure 4.19). The relationship of birds to tide corresponds well with the relationship of fish to tide (figure 4.7). The number of successful dives and the number of completed dives tended to vary with tidal stage and were greatest on the ebb (Kruskal-Wallis test, $\chi^2 = 6.28$, df = 3, $p < .1$, $\chi^2 = 6.40$, df = 3, $p < .1$). Birds hovered more on ebbing water ($\chi^2 = 5.52$, df = 3, $p < .1$). The percent of dives that were successful, the numbers of

---

Table 4.12. Interactions among tern diving success, wind, prey density, and prey abundance (Kendall's *tau*).

	Successful Dives			Unsuccessful Dives		
	tau	n =	*p* <	*tau*	n =	*p* <
Wind speed	−.07	268	.2	.04	257	.4
Prey density	.09	255	.04	.09	248	.05
Prey abundance	.07	255	.1	.10	248	.03
Prey density/wind speed	.11	232	.02	.08	232	.09
Prey abundance/wind speed	.10	232	.04	.10	232	.03

	Percent Successful		
	tau	n =	*p* <
Wind speed	−.12	222	.02
Prey density	.05	213	.26
Prey abundance	.02	213	.66
Prey density/wind speed	.09	202	.08
Prey abundance/wind speed	.04	202	.4

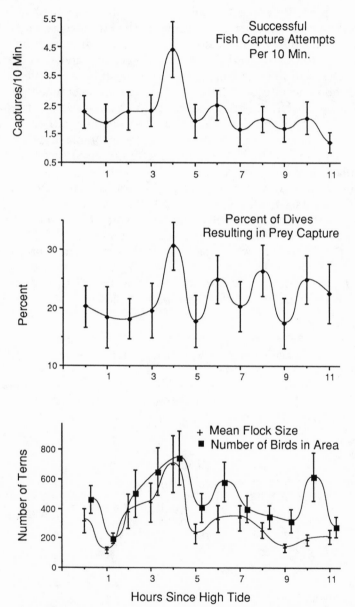

Figure 4.19. Tern diving success rates relative to tidal changes (means ± SE).

agonistic interactions, the rate at which birds left flocks, and the numbers of birds in the study area did not vary significantly among tidal stages ($\chi^2 < 5$, df = 3, $p \geq .2$). Flock size varied among tidal stages ($\chi^2 = 6.36$, df = 3, $p < .1$) and was highest during late ebb.

Fish Under Tern Flocks Versus Adjacent to Tern Flocks

Fish were generally more plentiful directly under foraging terns than in water adjacent to tern flocks. This difference lessened over the course of the season (figure 4.20). Although the contrasts were always statistically strong, the differences were greatest prior to chick fledging (table 4.13). This reflects the general decline of prey over the season, and its scarcity after the majority of young terns have fledged.

The Effect of Bluefish Presence on Prey Fish Under
and Adjacent to Tern Flocks
When differences in prey density, abundance, depth, and school size are analyzed separately for flock transects with and without bluefish, differences in the strength of contrasts in prey under versus adjacent to tern flocks tended to be stronger in transects where bluefish were absent (table 4.14).

When we contrast fish under versus adjacent to terns in flock

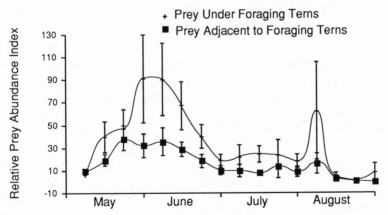

Figure 4.20. Changes in prey fish abundance under versus adjacent to foraging terns during summer (means ± SE).

transects with regard to both the terns' breeding cycle and the effect of bluefish presence, several patterns emerge (table 4.15). During the breeding season prey contrasts were strongest during the chick-rearing phase. Also, during the breeding season contrasts were always greater when bluefish were absent than when they were present. Postbreeding differences in prey fish density under versus adjacent to tern flocks were still greatest when bluefish were absent, but the contrast in prey abundance was greater when bluefish were present. This suggests that, as the season progresses and prey become scarcer, bluefish are increasingly important to terns in increasing the availability of prey and in facilitating the location of prey by terns. The strength of contrast of bluefish under versus adjacent to tern flocks decreased over the season. This may be because terns forage over bluefish only when bluefish are actively engaged in feeding, and as prey decline over the season a smaller proportion of the bluefish present are engaged with prey at any given time.

For all prey combined, density, abundance, depth, and school size are significantly different under versus adjacent to tern flocks (table 4.16). When sand eels and anchovies are examined separately, sand eel density and abundance are higher under tern flocks than adjacent to them, whereas anchovy density is higher under terns but anchovy abundance is not. This may relate to the fact that anchovies stay deeper on average than sand eels (Safina and Burger ms. a). An examination of the differences in predatory fish distribution under versus adjacent to foraging terns reveals that bluefish density was significantly higher under terns but that bluefish abundance, depth, and school size was not (table 4.17).

Table 4.13. Variability in the strength of contrasts between prey under flocks of foraging terns versus adjacent to them, with respect to date.

Date:		Before June 10 (incubation)	
	t	n	$p <$
Prey abundance	4.11	76	.0001
Prey density	6.25	68	.0001
Prey depth	−4.36	67	.0001
Prey school size	5.01	68	.0001

Comparison of Flock and Control Transects

Control transects were run in the same location, while flock transects were run through flocks of feeding terns, which could form anywhere in the study area. Flocks tended to form either at some physical feature that caused prey to come close to the surface, such as the inlet, sand bars, tide lines, or shoals, or in deep water where bluefish were active and were the cause of prey approaching the surface. We suspected that physical properties of flock and control transects might differ. We found no differences in water temperature or surface conditions between the transect types, but control transects had higher clarity on average (Kruskal-Wallis test, $\chi^2 = 5.47$, df $= 1$, $p < .02$).

Measures of fish differed strongly between flock and control transects (table 4.18). Values were greater in flock transects for all measures except fish depth, which was greater in control transects. Thus, fish were closer to the surface and more available to foraging terns in flock than in control transects. Comparisons of prey abundance between transect types is not very meaningful, because the abundance of fish in control transects is a function of the arbitrarily chosen fixed size of those transects. If the control transects had been twice as long, prey abundance would presumably have been twice what it was. In light of this, the difference in prey density is particularly striking (figure 4.21), especially because prey density and abundance are correlated within flock and control transects, as stated above. This emphasizes that for a given biomass of prey, concentrations were much higher in flock transects. Predator density was much higher in flock transects because bluefish schools

June 10 to July 15 (chick rearing)			After July 15 (postfledging)		
t	n	$p <$	t	n	$p <$
4.78	130	.0001	2.38	58	.02
5.67	104	.0001	4.22	43	.0001
−3.23	104	.002	−2.58	43	.01
3.62	104	.0005	3.07	43	.004

Table 4.14. Contrasts (t tests) between prey under flocks of foraging terns versus adjacent to them, with respect to bluefish presence.

	Without Bluefish			With Bluefish		
	t	n	$p <$	t	n	$p <$
Prey abundance	5.37	156	.0001	4.13	109	.0001
Prey density	8.34	156	.0001	4.53	109	.0001
Prey depth	−5.14	147	.0001	−3.07	70	.003
Prey school size	6.28	147	.0001	2.85	70	.006

Table 4.16. Contrasts of fish under versus adjacent to tern flocks for all prey combined, sandeels, and anchovies.

	Combined Prey	Sandeels	Anchovies
Density	9.33(268) ****	6.20(133) ****	5.28(42) ****
Abundance	6.47(268) ****	2.14(133) *	1.14(42)
Mean fish depth	−5.98(218) ****	−4.50(113) ****	−2.05(36) *
School size	6.63(219) ****	2.97(115) **	6.54(40) ****

NOTE: Numbers listed outside parentheses are values of t; those inside parentheses are n.
* = statistically significant at $p < .05$.
** = statistically significant at $p < .01$.
*** = statistically significant at $p < .001$.
**** = statistically significant at $p < .0001$.

Table 4.17. Contrasts (t test values) for predatory fish (Bluefish) under versus adjacent to tern flocks.

	t	$p <$
Bluefish		
Density	2.07	.04
Abundance	1.78	.08
School size	1.76	.08
Mean fish depth	0.87	.40

Table 4.15. Contrasts of prey and bluefish under versus adjacent to tern flocks, in transects with and without bluefish, during different tern breeding phases.

| | INCUBATION | | CHICK REARING | | POSTFLEDGING | |
	Bluefish	No Bluefish	Bluefish	No Bluefish	Bluefish	No Bluefish
Prey density	1.77[19]	6.55[56]****	3.51[54]***	4.36[74]****	2.50[36]*	3.60[22]**
Prey abundance	1.75[19]	3.84[56]***	3.42[54]***	3.96[74]***	3.10[36]**	2.01[22]
Predator density	2.37[19]*		1.61[54]		1.45[36]	

NOTE: Numbers listed outside parentheses are values of t; those inside parentheses are n.
* = statistically significant at $p < .05$.
** = statistically significant at $p < .01$.
*** = statistically significant at $p < .001$.
**** = statistically significant at $p < .0001$.

were highly mobile and the terns tracked them where they were feeding most actively.

Flock transects, despite their higher fish densities, reflected the same patterns of fish dynamics over time as control transects (figure 4.22). Thus it appears that the terns' method of foraging, while constrained by ambient trends of prey dynamics, allows them to evaluate a variety of patches and choose those with the highest concentrations at any given time. This appears not to be simply a function of their inability to locate fish at low densities, because we have previously shown that as ambient prey densities drop, the terns forage in areas where densities were lower than those that had previously been ignored (Safina and Burger 1985).

We compared the correlations between prey and date in flock and control transects to see if tracking of prey by terns mitigates the decrease in ambient prey. Although prey density, abundance, and depth were strongly inversely correlated with date in both types of transects, the relationship was even stronger in control than in flock transects (table 4.19), suggesting that tracking by terns does in fact mitigate the decline in ambient prey levels.

Table 4.18. Comparison of fish in Flock versus Control transects (Kruskal-Wallis test, df = 1).

	Transect Type with Highest Values	χ^2	$p <$
Prey fish			
Density	Flock	100.49	.0001
Abundance	—	0.47	.5
School size	Flock	69.20	.0001
Mean depth	Control	75.15	.0001
Predatory fish			
Density	Flock	26.58	.0001
Abundance	Flock	12.54	.0004
School size	Flock	22.29	.0001
Mean depth	Control	30.54	.0001
Feeding activity	Flock	61.37	.0001
Feeding intensity	Flock	52.58	.000

■ Discussion

The ocean is not simply a vast volume of water in which and over which animals roam in search of prey patches that swim and drift at random. Like the land, it is a complex mosaic of habitat types. Unlike the land, where habitat patches are rather fixed and change very slowly within our perception of time, ocean habitats are constantly changing, moving, and hidden from our view. They are

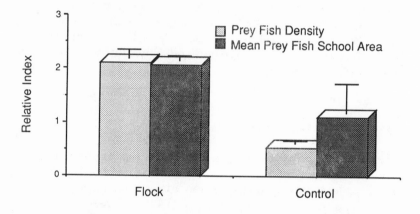

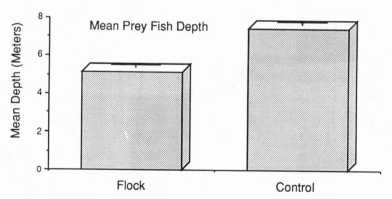

Figure 4.21. Comparisons of prey fish depth, density, and school size between flock and control transects (means ± SE).

delineated by currents, mixing patterns, thermoclines, surface temperature fronts, light penetration, and upwelling caused by the interaction of currents and ocean floor irregularities. This is what dictates the distribution of marine organisms. To a terrestrial animal, the earth must seem like a patchwork of habitat types, out of which it chooses where to look for required resources based on its particular species' needs and a combination of neurologically innate tendencies and learned preferences. To a fish or seabird, the ocean, too, must seem a patchwork, but here the patches are constantly in dynamic motion, and the challenge to the motile inhabitants is to track the habitat patch, to locate other suitable patches, and perhaps to anticipate the formation of favorable patches. The challenge for us is to try to understand and share the marine animal's oceanic map—to know where animals are and why, not only to describe their movements but to account for them.

Physical Variables and Fish

Though the behavioral responses of ocean fish to physical factors such as tides and wind are well known by fishermen, who pay considerable attention to them, the subject has been given little formal consideration by scientists (D. Conover, personal communication). This seems due in large part to the fact that ocean fisheries researchers are mostly concerned with population assessment

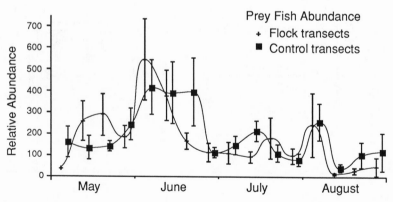

Figure 4.22. Comparison of prey fish abundance between flock and control transects (means ± SE).

(i.e., providing information on the effects of fishing), whereas fishermen are concerned with behavior (where and when will fish be catchable). Illustrative of the paucity of information on the effect of physical variables on fish behavior, Levinton (1982) in his recent classic text *Marine Ecology* reviews tides and wind as physical phenomena that affect water mass circulation, but information on how they affect biological variables is lacking. For comparison, Lyman and Woolner (1954), in *Striped Bass Fishing,* anecdotally mention tides, wind, and weather, in relation to how to find and catch striped bass, on 41 pages of their 230-page book. Clearly, physical variables and their effect on fish behavior have been important to fishermen for quite a while. Where discussions of tides and fish behavior have entered the scientific literature, they have usually been in relation to fishing. Richards (1976) noted that tide had an effect on the rate at which bluefish were caught on hook and line during sampling. Madsen (1963) discussed tides with respect to sand eel catches.

Clarity and Temperature
   In our study, water clarity and temperature differed from year to year. Such differences appeared to be due to stochastic variations in wind regimes and their effect on mixing. In our area, southerly winds blow warm water inshore and allow it to stay near the surface where it is further warmed by the summer sun. These onshore winds also aid the formation of groundswells and rough surf which increase the suspension of sediments and lowers clarity. Northerly winds have the opposite effect, blowing warm surface water off-

Table 4.19. Correlations of date with prey variables in Flock and Control transects.

	Flock			Control		
	tau	n	$p <$	*tau*	n	$p <$
Prey density	−.14	278	.0007	−.24	263	.00001
Prey abundance	−.16	278	.0001	−.25	263	.00001
Prey depth	−.31	250	.0001	−.27	237	.0001
School size	−.07	250	.09	−.11	237	.01

shore, allowing it to be replaced by cooler deeper water, and flattening the surface near shore, allowing settling of sediments and subsequent increases in clarity. In 1984, for example, an unusual and virtually constant regime of very light northwesterly winds for a week and a half caused the water to become exceptionally clear.

Straty and Haight (1979), discussing the Bering Sea, note that water temperature can vary significantly from year to year. They briefly mention several possible effects in colder water, including 1) a potentially increased vulnerability of fish to bird predation due to reduced swimming speed and slowed growth (increased time at vulnerable sizes) of fish, and 2) reduced competition between seabirds and larger fish due to reduced foraging activity of larger fish. Warm water, they posit, would have the opposite effects. It seems doubtful that in the system we studied the temperature changes would be great enough to have these effects. Even if it did, these effects would not necessarily work in concert. For example, a decrease in temperature great enough to affect the swimming speed and feeding activity of fish would be likely to decrease their vulnerability to plunge divers, because prey fish are most vulnerable when they are near the surface feeding or fleeing feeding predatory fish. We would predict that the range of interannual temperature variations in our area work more subtly, primarily by affecting fish movements. We did not obtain enough data to make declarative inferences, but the data may suggest that warmer water brings both prey and predatory fish earlier. The phenology of bluefish migration, at least, appears influenced by temperature (Lund and Maltezos 1970).

Prey and predatory fish depth increased both with increasing water depth and also with increasing water clarity. It has been our impression that prey fish tend to avoid the surface if possible and approach it only when necessary to feed on plankton or escape piscine predators. Fish may routinely avoid the surface at other times. At times when few or no birds were present, fish seemed to remain away from the surface. At other times, the tendency may have been enhanced by a response to birds in our transects. Fish are known to respond evasively to the sight of flying predators overhead (Milinski 1984; Romero 1985), and this can affect fish distribution (Power 1984) and, presumably, the evolution of seabird coloration (Cowan 1972). If seabird predation applies significant pressure to the prey community, as several estimates suggest (e.g., review by Furness 1982; Bourne 1983), then we would expect prey fish to exhibit a suite of behaviors that decrease their risk to birds.

The Influence of Water Depth

Within our area, predatory fish abundance did not change with depth, but prey abundance was greater in deeper water. There are at least three possible reasons for this: 1) Over the course of the season, prey fish moved inshore and also declined in number. The greater abundance in deeper water may relate more to the fact that they were there earlier in the season, when they were more abundant, than to depth per se. 2) Prey tended to avoid both the surface, where birds threatened, and the bottom, where bluefish lurked. Deeper water allowed maximization of distance between surface and bottom, possibly attracting numbers of fish. This effect might have been countered by the fact that arriving bluefish came inshore from deeper waters, accounting for the progressive shift by prey into shallower water over the season. 3) Prey may be more at risk from predators when in deeper water and tend to form larger groups there. The coalescence of small groups of sand eels as they moved into deeper water was described by Kuhlmann and Karst (1967). Predator abundance was not related to water depth, though predator size was. It is conceivable that larger predators might be more efficient, pose more of a threat, and prompt a fuller expression of facultative prey defense (schooling) than smaller predators. On the other hand, prey density and prey school size, which are related to bluefish presence (Safina and Burger ms. b), did not show a clear relationship with depth. Overall, then, we have little support for the idea that group defense, *sensu* Hamilton (1971), was involved in making prey more abundant in deeper waters.

Our study area did not extend out to the deep ocean, and most of our data was collected in water less than 15 meters deep. Because we worked only in relatively shallow ocean depths, it is possible that correlations of fish with water depth may not be accurately extrapolated into general rules of fish/depth relationships.

Tidal Effects on Fish

Fish responded measurably to tide only in flock transects. Control transects were in close proximity to the inlet mouth, but were out of the main current stream. Flock transects often occurred in or just outside the inlet mouth in areas of relatively high current velocity. Prey fish activity generally was higher on the ebb, presumably because prey fish gathered in response to food being swept out of the estuary. Scott (1973) noted that sand eel catches were best during the part of the tide that made the most food available.

Predators responded later in the tidal cycle, in apparent response to the aggregation of prey.

Fish Responses to Weather and Sea Surface Conditions

Windiness generally seemed to "scatter" prey fish in control transects, as indicated by its effect on abundance, density, and our school size index. However, prey fish were nonetheless tracked by terns in flock transects, where wind speed (up to the moderate conditions in which we worked, to approximately 24 km/hr) had little effect on fish. Dunn (1973) found that an increase in wind speed up to moderate conditions enhanced tern foraging. It is possible that the increase of bluefish in control transects in response to higher wind speed, which we suspect relates to onshore pushing of warmer water by southerly winds (these tend to be the stronger winds during summer), contributed more to the changes in prey fish than did wind per se. Fishermen casting for bluefish from shore generally learn to expect better success when winds are onshore.

Cloud cover's effect on fish was limited to bluefish in control transects. Bluefish were more likely to be present in control transects as cloud cover increased. In this regard the effect resembled that of wind, especially if the mechanism causing the response of prey to wind was wind's effect on bluefish movements toward shore. Southerly winds are also those that increase cloud cover, so the similarity in effect may be due in part to common origin of the phenomena.

Bluefish presence, abundance, and density were more depressed by surface choppiness in flock than in control transects, while their feeding activity was more depressed in control than flock transects. This is perhaps an artifact of the relationship between bluefish feeding and the tracking of bluefish by terns. Nonfeeding bluefish are not likely to appear in flock transects because terns do not usually congregate over them. It is reasonable to hypothesize that bluefish do not track prey when they are not feeding. Feeding terns, of course, always track prey. Thus we may expect (as observed) bluefish to be absent from flock transects when they are not feeding. We may further expect that the tracking of actively feeding bluefish by terns would, to some extent, mask the bluefish's tendency not to feed when the water surface was rough, because terns would preferentially track only those bluefish that were actively feeding, even if those fish were a fraction of the local population (in which case measures of their density and abundance in transects would be low) and/or they did not feed for long. Such masking would produce an apparently stronger response in numbers in flock than in control transects, while measures of their activity would be greater in controls, as we observed.

The apparent responses of fish to water temperature must be considered cautiously. Because temperature correlates with date, it is difficult to assess the relative importance of water temperature, date, and fish predator-prey interactions in causing the observed patterns. Fish variables were more strongly correlated with date and predator-prey interactions than with water temperature. In particular, the fact that mean prey depth was related to temperature in flock but not control transects may have more to do with the increasing prevalence of bluefish in flock transects as the summer progressed than with any other factor. It may be, however, that the limited range of depth in control transects (water depth there was 6–12 m) partially masks any tendency of fish to change depth. In any case, water temperature probably does influence the arrival and presence of prey and predatory fish (Lund and Maltezos 1970).

Bird/Fish Relationships

Size and Density of Bird Groups

Bird numbers correlated with prey abundance and prey density, especially within the upper 3 meters of the water column. Bird numbers also correlated with prey school size. There was no linear relationship between bird numbers and mean prey depth. Prey depth interacts with abundance in a way that affects availability to terns; dividing prey abundance by mean prey depth resulted in a closer correlation with bird numbers than did prey abundance alone. Bird numbers were also related to predatory activities of bluefish. The attractiveness of bluefish to birds relates largely to the bluefish's effect of driving prey to the surface (Safina and Burger ms. b; figure 4.23), and bluefish presence enhanced the attractiveness of a given quantity of prey to terns.

In qualitative terms, peak seabird densities generally occur in areas of the ocean containing relatively concentrated food (Obst 1985; Abrams and Underhill 1986). But, as noted by Woodby (1984), attempts to quantitatively understand bird distribution in terms of prey density have often been inconclusive or shown inconsistent relationships. In Woodby's (1984) work, murres were not found in the densest food patches. Woodby speculated that murres may feed adequately in the less dense patches and that weather might separate them from dense patches, inhibiting a consistent relationship. Briggs et al. (1981) reported that the distribution of the anchovies that were the brown pelican's principal prey did not correlate well with bird densities, and that the correlation was better in the spring than in the fall. Braune and Gaskin (1982) noted that the distribution of larids followed prey distribution. Obst (1985) found for Antarctic waters that seabird density was a good predictor of krill presence but that bird density did not correlate with krill density or depth. In a paper describing our first year's work on this study (Safina and Burger 1985), we noted that the presence of terns was a good indicator of dense prey. Briggs et al. (1981) noted that prey abundance and availability may be very separate things, and that seabirds may respond to different perceptions of food "maps" than the ones we obtain. This seems likely to be even more important in deep ocean areas than in nearshore shelf areas such as the one we worked in.

Figure 4.23. Feeding frenzy of terns over bluefish (note splash patterns).

Obst (1985) points out that problems in relating bird density patterns to food density distribution may be dependent on the scale at which data are collected, and that, within the broad limits set by the physical ocean, small-scale phenomena might be very important. Having worked on a relatively fine scale, we concur, and would add that the temporal scale of data collection is important as well. Compared to single-season studies, longer-term work yields better intuitive and empirical understanding of systems that change dynamically within and among years, and of relationships that are inherently imperfect but that have profound biological importance, such as a seabird's knowledge of what lies out of sight.

It is the density of available prey, which for our plunge-diving terns means prey near the surface, that showed the best relationship with flock size. This is not surprising, because prey deeper than about 0.5 meters must be largely unavailable. We routinely watched terns coursing over prey that were too deep; occasional dives were made on prey that came near the surface. Our direct observations of prey in these situations indicated that, although the main body of the school would remain out of reach (often 3–5 m below the surface), individuals were leaving the dense school to enter surface swarms of plankton, ingesting prey, and then retreating to the safety of the school. Individual prey fish sometimes seemed to rise from the school to a particular prey item, like a trout rising to a mayfly. If individual fish were in fact responding to individual prey, there would be no reason for the whole school to rise together to take advantage of group defense, because there would not be an equal opportunity for "group profit" among all individuals. We believe that having most of the prey school remain away from the surface is a strategy that minimizes the risk of avian predation for most fish most of the time. Paradoxically, this is how individuals were most at risk (excluding occasions when they were driven to the surface by predators). Terns could achieve high capture rates at these times by singling out the rising individuals. Milinski (1984) showed for sticklebacks that attentiveness to invertebrate prey was inversely related to attentiveness to avian predators. Balancing the demands of hunger and the risk of predation is fairly common in nature (Sih 1980), and animals must often leave secure refuges to search for sustenance. In those relatively rarer situations when the entire school surfaced to feed, something that bay anchovies occasionally did, terns did not dive as much as one might expect. Predators in many taxa have great difficulty catching swarming prey (Wilson 1975; Milinski 1984), and this seems especially true when the prey are capable of swift evasion. Grouping by prey is generally believed to have evolved as a defense against predators (Hamilton 1971). In our experience, avian predators such as terns and raptors often do not even initiate attacks on grouped prey, and in the case of raptors at least, the inhibition of attacks on grouped prey is progressively reinforced through failure, apparently through confusion, during the ontogeny of hunting behavior and the development of skills in young individuals (Safina, personal observation).

When prey leave a school to approach the surface, the amount of prey that come within striking range of terns seems to be directly related to the amount that are out of range (too deep). Because prey density and abundance are greater when prey is deeper, we might expect more prey to appear near the surface when prey average deeper than we would expect if depth of prey was not related to prey numbers. This would obviate a direct relationship between mean prey depth and prey availability, and may be partly responsible for our observation that flock size and numbers of foraging terns did not respond to changes in mean prey depth.

Defense of Feeding Space

Tern density was inversely correlated to prey abundance, indicating that terns prefer to spread out when they can. If we may extend Hamilton's (1971) argument from grouped prey to grouped predators, the amount of prey in a selfish predator's "domain of availability" is proprotional to the distance to the nearest neighboring predator, if prey is distributed evenly within a prey patch. This is, of course, the basis for the phenomena of food territoriality. In fact, terns sometimes defended ephemeral foraging territories within foraging flocks. This previously unreported behavior differs from that described by Nisbet (1983) in that territories were not related to geography and lasted only as long as the bird maintained its position in the flock, usually on the order of minutes. In our observations territoriality within flocks seemed most likely to occur when food was present but not readily available, catch rates were low, predatory fish were absent, and terns were not very dense. Typically, a relatively large, deep school of prey would be present, as often occurred early in the season. Territoriality was manifest by birds coursing in ellipses approximately 50 meters long within the flock, but maintaining, to a large extent, exclusive use of the area within the ellipse by vocal threats, minor movements toward intruders, and occasional chases and fights. Visually, the effect was subtle; any pattern of order was largely obscured by terns entering and leaving the area or traveling through territories, and by the fact that not all terns were defending an area.

Indefensibility of feeding territories for species whose prey is ephemeral and patchily distributed has been cited as one of the primary reasons for the evolution of coloniality in birds (Horn 1968; Wilson 1975: 52–53; Burger 1981). It is generally assumed that this is an all-or-nothing process; either the prey is defensible, and feeding territories form, or it isn't and they don't. In fact, however, a continuum would seem to be a more realistic description (at least for common terns); prey are defended to the extent that they are defensible and, presumably, to the extent that the energy return exceeds that spent on defense (Carpenter and MacMillen 1976). Nisbet (1983) showed that the defense of shoreline territories by common terns was economically very favorable where it occurred. In certain places and at certain times, then, either feeding territories are established along particular reaches of shoreline and defended repeatedly by birds breeding in distant colonies (Nisbet 1983), or terns defend small areas around themselves in flocks for very short periods of time, or they do not defend a foraging area, depending on circumstances.

Advantages and Disadvantages of Flocking

Foraging in flocks can serve at least two functions: increased foraging efficiency and decreased predation risk (Powell 1985). Where prey is both scarce and patchy, foraging in flocks is beneficial, both by increasing the food intake rate and decreasing the variation in the rate, even if the group serves no antipredator function (Clark and Mangel 1984). However, past a certain density, the competitive effects of flocking can weigh against the benefits, and Clark and Mangel (1984) and Sibly (1984) have shown that socially optimal flocks (in terms of rate of food intake) are unstable, and that larger-than-optimal flocks represent an evolutionarily stable strategy. Sibly's model estimates that flocks can be expected to be twice the optimal size. It is interesting to note that group sizes are, at least sometimes, larger than optimal even in cooperatively hunting animals such as African lions, although other ecological factors may be germane in this case (Caraco and Wolf 1975). With our terns, the highest success rates were achieved in midsized flocks (figure 4.24), but mean flock size (297 ± 26) coincided with high fish capture rates.

Within prey patches, tern density (number of terns divided by transect length) was inversely related to mean prey fish depth and positively correlated with the prey density in the upper 3 meters, but not with the density of deeper fish. Tern density responded strongly and positively to predatory fish that were feeding, and here the visual effect was less subtle. If many bluefish were exploding through the surface in pursuit of showering numbers of prey, the sight and sound was spectacular. Here, too, terns attempted to decrease interference competition from their neighbors with much vocal threatening, but the area that could be defended was limited to the immediate area around the individual, because birds were often dense enough to collide in the intensity of their excitement. In these situations, prey availability and its location at the surface constantly changed. Prey fish appeared and disappeared seemingly at random, depending on the movements of bluefish below, and tern groups shifted frequently. Even while hovering close to the water, groups of terns often moved along over the water's surface as though tracking individual or small pods of bluefish. The sudden appearance of prey prompted rapid diving (figure 4.25), when only the closest terns of a densely hovering mob had a good chance of taking a fish (similarly noted by Duffy 1986). Terns frequently did not need to enter the water at these times, but could arc down and seize prey from the surface while in flight. Prey was sometimes taken while it was leaping to avoid a bluefish. Small and seemingly fragmented bluefish schools working within a larger area appeared and disappeared unpredictably at times, and terns searching individually quickly coalesced where one tern stopped to hover or dive,

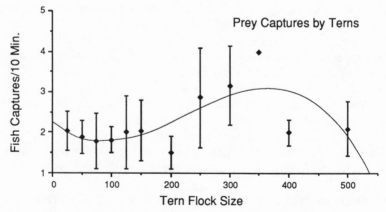

Figure 4.24. Mean fishing success (± SE) of terns relative to size of foraging flock.

often traveling rapidly hundreds of meters to do so (similarly noted by Gochfeld and Burger 1982). Often only a small fraction of the birds would reach the spot before the availability of prey subsided, whereupon all birds would begin searching anew.

Tern Diving and Prey Fish

The diving activities of terns indicated that, although the diving rate was related to both density and abundance of prey, the density distribution of prey is more important to terns than is the biomass of prey. The importance of the distinction between prey abundance and prey availability has been made before (e.g., within a theoretical context by Charnov, Orions, and Hyatt 1976, and within an empirical context by Briggs et al. 1981). Not surprisingly, prey density and prey depth were related to the prey capture rate of terns; higher densities of prey within 3 meters of the surface resulted in increased capture rates. The exception appears to be dense swarms of bay anchovies at the surface, as noted earlier. The relationship between tern success rate and prey density diminished as prey densities at increasing discrete depths from the surface were considered.

Although terns may form flocks over deep prey, they cannot always depend on deep prey coming to the surface. If the water is

Figure 4.25. Closeup of terns diving for fish over a bluefish school.

clear enough, terns can see and sometimes dive for prey which is normally out of reach, as indicated by the fact that the number of unsuccessful dives increased as mean prey depth increased. Flocks sometimes gathered over deep prey with little success. On some occasions this seemed due to a dearth of alternative patches, and on other occasions it seemed that terns gathered in anticipation of a more favorable stage of tide when more prey came to the surface and catch rate increased.

Whether or not the terns really anticipated increased prey capture rates in these situations is difficult to say. Braune and Gaskin (1982) found that numbers of nonfeeding gulls and terns in an area peaked one tidal phase before numbers of feeding birds peaked, and that "anticipation of prey availability is strongly suspected." Anticipation clearly plays a role in their foraging at times. In Long Island Sound, we have observed terns arrive at a productive shoal 5 kilometers from their breeding colony in deep darkness just prior to the first light of morning. In our study area, terns did not disperse randomly from the colony at first light and search until prey was located, but, rather, the bulk of the morning's foragers flew directly to areas that had been productive the day before or that were usually more productive than other areas. Thus, they have a foraging success memory that enables them to reduce searching time and may help them to evaluate the relative quality of patches by providing them with an anticipated acceptable success rate upon which to base patch leaving decisions. The interesting problem of how animals derive expectation is yet to be developed (Krebs, Houston, and Charnov 1981).

We found that the relationships between prey presence (density and abundance) and tern diving grew progressively stronger as we considered numbers of successful, unsuccessful, and aborted dives in turn. This suggests that the ratio of each "type" of dive, within the total number of dives initiated, changes dynamically with changing fish parameters and is not simply linear. The fact that only a weak correlation was found between prey density and successful dives, while stronger correlations were found between prey density and unsuccessful and aborted dives, may suggest that terns are less careful when food is plentiful.

Animals may waste energy, or may make decisions that result in lowered profits, in situations where the consequences of those decisions are small. Sih (1982) states that "we might actually expect" foragers to be relatively inefficient when the feeding rate does not strongly affect fitness. Hodges (1985) found that bumblebees followed a threshold departure rule in choosing to leave or stay in foraging patches, and that this rule was resilient to changes in patterns of dispersion of flowers and plants, but that the rule was violated when the cost of violation was slight. Further, wastefulness of energy can actually be profitable, if the food intake rate over time is improved by spending less time on each individual food item in favor of handling more items. For example, Bailey (1986) found that predatory insects were more wasteful at high food densities, discarding prey after extracting only 60 percent of the available food. However, this allowed them to feed for less than half the time on each prey item, giving them more hunting time to respond profitably to the high density of ambient prey. If terns increased the rate of dives when fish were plentiful by making more careless dives (i.e., not taking time to aim well), they could increase their overall rate of prey intake even if the cost of "hasty" dives is a lowered percentage of success per dive. In human business interactions, lowered rates of return per "prey individual" (customer) in favor of encountering more prey individuals (attracting more customers) is precisely the strategy behind discount retail stores, which appear in urban areas where "prey" is dense and operate under a policy of reduced profit per item sold in order to sell more items to more customers and thereby increase net currency gain.

Agonistic interactions, which were mainly about spacing (rather than piracy), increased in frequency as measures of prey availability decreased. An increased prey depth, decreased capture rate, increased rate of aborted dives, and increased sea surface choppiness corresponded with increasing rates of agonistic interactions. Because aggression was largely used to increase or maintain individual distance while foraging, we view its relationship to food as that of latent tendencies toward food territoriality becoming manifest under certain conditions. For birds and other animals in terrestrial habitats, the degree of territoriality and size of territory can vary inversely with food supply (e.g., Laskey 1935; Stenger 1958; Simon 1975; Zach and Falls 1976; Lederer 1977). Birds that are strongly territorial under normal conditions may cease to behave territorially in the presence of superabundant food (Michener 1951). Alternatively, aggression during food scarcity may in part be displacement (Tinbergen 1952) facilitated by frustration (Miller 1941).

Tern Diving and Predatory Fish: Competition or Commensalism?
Terns and bluefish have a trophically well-integrated relationship. Not only do terns eat juvenile bluefish (figure 4.26), but adult

bluefish apparently eat terns on occasion (we have seen capture attempts, and several fishermen have observed captures). On more than one occasion we have seen bluefish leap into the air to seize a lure that had not quite reached the water. Bluefish, much noted for voracity, may also attack other water birds (French 1981).

Despite the very strong attractiveness of bluefish schools to terns, terns initiated fewer dives over feeding bluefish. We believe that this is because bluefish reduce the overall numbers of prey in an area over time, forcing the terns to rely more and more heavily on feeding bluefish to concentrate scarce and dwindling prey. Because they consume large quantities of prey and appear to induce a general flight of prey out of an area, the presence and activities of bluefish have a strong depressive (*sensu* Charnov, Orions, and Hyatt 1976) effect on prey; they appear to be the primary trophic competitors of terns in the New York Bight and the main factor causing the seasonal decline in prey fish numbers and availability to terns during the summer (Safina and Burger 1985 ms. a and b).

Seasonal declines in food availability are a dominant feature in the evolution of many communities (e.g., Kushlan 1981; Schluter

Figure 4.26. Adult common tern leaving flock with a fish to take back to the colony to feed young.

1982). Where seabird populations are large, they may predate a significant portion of the seasonal fish production (review in Furness 1982; Whoriskey and Fitzgerald 1985), although estimates of their effect vary greatly (Bourne 1983). Thus the terns may conceivably be a factor in the seasonal decline we observed. However, even if the effect of seabird predation on prey populations is substantial, and even if food supply limits many seabird populations (Ashmole 1963; Furness 1982; Bourne 1983; Schreiber and Schreiber 1984), this still does not imply that seabirds have more of an effect on the behavior or numbers of prey in a system than do predatory fish. This seems especially true of birds that do not pursue prey below the surface.

The effect of bluefish on prey numbers and prey behavior is very strong and very rapid (Safina and Burger ms. b); we feel that it is the overwhelmingly dominant predatory effect on the prey fish populations in the system we studied. Bluefish and terns occupy the same consumer level in the marine food chain. Their diets overlap greatly during the period of the year while they are in shared feeding ranges, and their habit of feeding in the same place at the same time, often chasing the same prey individual at the surface, puts them in direct competition. Yet there is, of course, a fundamental difference in their distribution and habitat. The food of plunge-diving seabirds such as terns exists in a different and hostile environment, within which the bird cannot survive more than the few moments its breath will last, and into which the bird cannot penetrate more than a few tens of centimeters (perhaps 40 cm). Thus, despite their dietary similarity, bluefish and terns have very unequal access to prey, and the bluefish have a distinct advantage. Any prey within striking range of a tern is within range of a bluefish, except for the rare instances when bluefish chase prey into extreme shallows or even onto shore. But bluefish potentially have access to prey throughout the entire water column, while terns must wait until prey is near the surface. Because of this asymmetry, we view bluefish as the dominant predator in this system and the terns only as ecotonal species, dependent on the movement of the prey fish but unable to maneuver them into advantageous positions. Thus the terns are dependent variables with respect to prey fish, while prey fish are independent with respect to the terns. Although the terns do exert some pressure on the fish population, the terns' impact on prey numbers and behavior is probably minor compared with the effect of bluefish. The contrast between the continuing buildup of prey concurrent with the buildup of tern numbers early in the season, and the rapid decline of prey after the arrival of bluefish, is evidence for this.

In competitive interactions, the competitor with the greater access to food has a significant advantage. Where fish compete with birds for food, fish usually limit food availability to birds, and, ultimately, bird numbers, by more efficiently exploiting the common prey, thus reducing food availability to birds (Zaret and Paine 1973; Pehrsson 1984; Hunter et al. 1986). Most investigations of such interactions have been done with ducks on lakes. Because of trophic competition, duck abundance was inversely related to fish density (Eadie and Keast 1982), and ducks grew better and found food more easily in lakes without fish (Hunter et al. 1986). Ducks increased their use of lakes after fish were removed (Eriksson 1979). The effects of competition from fish on ducks is proportional to the niche overlap between fish and duck species (Eriksson 1983). When a predatory fish was introduced into a tropical lake, fish-eating birds (terns, kingfishers, and herons) as well as native predatory fish (tarpon) declined in those areas of the lake with the new predator (Zaret and Paine 1973).

The only advantage terns have over bluefish in gaining access to prey is that they can change locations much more rapidly—and they make use of this advantage. In 1985, for example, during a period of food scarcity which appeared to be bluefish induced, color-marked terns from Cedar Beach flew overland to the North Shore of Long Island (approximately 15 mi or 25 km) to forage on sand eel schools which were very dense in Long Island Sound at the time (J. DeBellas, personal communication). Terns have been doing this for a long time to varying degrees within and among years (Raynor 1972). This is another way in which foraging success can be buffered against reductions in food availability.

Despite their depressive effect on prey, the presence of bluefish in some ways enhanced the availability of prey to terns. Bluefish fractured prey schools, drove prey to the surface, and distracted prey fish attention from avian predators, thereby greatly diminishing the effectiveness of prey defense against tern predation. For the terns, this was manifest in fewer aborted dives and a higher percentage of successful dives when bluefish were feeding. Though the presence of bluefish did not generally result in a greater rate of prey capture over time, their presence did appear to enhance prey capture at low to intermediate prey densities (figure 4.12). And the presence of bluefish enhanced the availability of prey to terns through its effect on the interaction of prey density/prey depth. Further, terns could obtain a fish with less energy expenditure per dive because they often did not need to enter the water to seize a fish.

Terns hovered most when bluefish were feeding actively and were plentiful and near the surface. Aggression among terns was high when terns were doing much hovering, but aggression was not related to bluefish variables per se. Terns hovered most when they were excitedly trying to maintain preferable positions in dense flocks while tracking bluefish. Thus the aggression was directly related to the social competitiveness of the situation.

The presence of bluefish altered the relationship between prey distribution and prey availability to terns. In the presence of bluefish, the success of terns was related more to prey abundance than density. When bluefish were absent, the relationship was reversed, with success related to density, but not abundance. Prey abundance is generally low in the presence of bluefish (Safina and Burger ms. b). Bluefish concentrate prey by effectively herding it to the surface. Coming to the surface and leaping is an almost universal predator avoidance pattern in schooling fishes when they are attacked by schooling predators, which usually approach from below where they can see prey more easily and be seen less easily (Colblentz 1985). Leaping presumably causes the predator to lose continuous visual contact with the prey (Colblentz 1985). This is no doubt true when prey are grouped, but we often observed a singled-out fish being closely pursued through a series of leaps by a bluefish following just below the surface.

We had the impression that in water more than approximately 10 meters deep, terns were heavily reliant on bluefish to make prey available. Tern reliance on predators in deep water has often been commented on (e.g., Ashmole and Ashmole 1967; Harrison, Hida, and Seki 1983). Published work on interactions between seabirds and schooling predatory fishes has been largely anecdotal and decidedly bird oriented, describing predatory fish facilitating birds' feeding by driving small fish that would otherwise have been unavailable to the surface (e.g., Ashmole 1963; Ashmole and Ashmole 1967; Gould 1974; Brown 1980; Au and Pitman 1986). Most have presumed the relationship to be simply commensal, and have not looked into the population and behavioral dynamics among the fish. However, our studies indicate that predatory fish can have a profound effect on prey numbers and behavior, and competition may be a very important and previously overlooked aspect of this relationship (Safina and Burger 1985 ms. b). Bluefish and tern diets overlap greatly while they are both feeding in the study area, and their habit of feeding in the same place at the same time, often chasing the same prey individual at the surface, frequently puts them in direct competition. At other times, bluefish and terns compete through the effect that bluefish have on the terns' food supply.

Tern Foraging and the Physical Environment

We have discussed how the physical environment affects fish and how the distribution of fish affects birds. With this understanding, we can now look at how the physical environment affects birds through its effects both on fish and directly upon the terns themselves.

As water clarity increases, fish are visible farther from the surface (Eriksson 1985). This does not lead to increased captures, however. Birds did less diving, and aborted more of the dives they initiated, as water clarity increased. This is no doubt due to the tendency of fish to go deeper as clarity increases, and may be due to a direct response of fish to the sight of avian predators. Dunn (1973) thought that conditions that obscure a fish's view of the terns (moderately choppy surface conditions) did more to facilitate fish capture than to inhibit fish detection by birds. Thus, in plunge-diving terns, the prey's response to the physical conditions and to the predator dominates the encounter. According to Sih (1984), when prey are mobile and a spatial refuge exists, the prey response should dominate. For prey attempting to avoid common terns, depth can be the surest refuge. On the other hand, when prey mobility is restricted by the habitat and refugia are unavailable or poor in quality, avian predators may inflict heavy mortality. For instance, during dry periods when prey in ponds are concentrated by low water levels which allow access by wading birds, herons can reduce both fish biomass and numbers by more than 75 percent (Kushlan 1976).

Terns did less foraging as time of day increased toward midday, and their success rate declined. This may relate to prey becoming less available as light penetration increases because prey can better see and avoid terns. Prey depth did not increase with time of day in flock transects, nor did prey density, but prey abundance was reduced over time during the morning if bluefish were not present. It is possible that the reduced prey abundance is related to a majority of fish disengaging themselves from tern tracking, possibly by increasing their depth or, for sand eels, burrowing. Disengaged prey would not be measured in flock transects, and control transects offered only limited range of depth and so cannot help answer this question. It seems likely that feeding bluefish could facilitate tern tracking, resulting in little disengagement of prey from terns and little measured difference in prey abundance, as we observed.

Differences in clarity and cloud cover had no effect on the number of birds present or the size of their flocks, but tern flocks were denser when cloud cover was greater and water was less clear. Conditions associated with high light penetration into the water, then, resulted in more dispersal of the same number of birds. The dispersal of feeding birds in a prey patch can reduce interference from neighbors (Scott 1984). Dunn (1973) did not find a relationship between tern foraging success and cloud cover, nor did we. Glare, apparently important in inhibiting feeding for wading birds (Bovino and Burtt 1979), may not be as important for terns.

Wind speed was positively related to tern density but not to numbers of foraging birds. If wind decreases visibility through the surface, terns may group more closely to other foragers because searching terns experience increased difficulty locating prey on their own. Wind speed did not affect prey capture rates per unit time, but it was inversely correlated to the percent of completed dives that were successful. This may mean that wind can upset aim and trajectory (Dunn 1973). Although wind speed can be related to sea surface conditions, depending on the wind's direction, sea conditions did not correlate to the diving or prey capture rates of terns in our study. Dunn (1973) found that both prey capture rates and dive success were affected by wind speed, and that prey capture was enhanced by wind, at least up to moderate wind conditions. Sea conditions did affect prey capture in Dunn's study, independent of wind; at any given wind speed, prey capture was better in moderate rather than calm conditions, probably because the tern was less visible to the prey fish.

Because clarity, cloudiness, and wind speed did not affect the numbers of foraging birds in our study but influenced their density, our results suggest that social attraction becomes more important as prey becomes more difficult to see. Birds often headed directly to foraging flocks, doing little searching on the way. It is easy to imagine that as conditions for locating prey deteriorate, individuals rely more on finding birds that have already found fish. Social attraction is an important phenomenon for colonially breeding birds that forage in flocks. Although the idea that colonies act as food

information centers (Ward and Zahavi 1973) has been contested
(Pratt 1980; Andersson, Gotmark, and Wiklund 1981; Evans 1982;
Erwin 1983b; Bayer 1982), the attractiveness of foraging flocks away
from the colony has been repeatedly confirmed (e.g., Gochfeld 1978;
Pratt 1980; Andersson, Gotmark, and Wiklund 1981; Gochfeld and
Burger 1982; Erwin 1983b; Bayer 1982). Gochfeld (1978) discussed
"feeding lines" of birds returning with fish, and the presumed ease
with which birds leaving the colony could fly against the flow of
such lines until they found the source. We confirmed this. In those
cases where lines of commuting terns formed, the source could
often easily be found, even if such lines were relatively sparse or
diffuse, and the feeding flock was out of sight miles from land, and
regardless of whether commuter lines were intercepted just off the
colony or well out on the ocean.

Falling tides were important to terns in our area. The effect of
ebbing water on fish and on food availability for both prey and
predatory fish, as discussed above, extends to the terns. Tidal effects
may be nested within time-of-day effects (Galusha and Amlander
1978). Our terns showed such a pattern. Tide has been found to be
a factor in the feeding patterns of a diverse group of water birds;
terns (Dunn 1973; Becker, Finck, and Anlauf 1985; Braune and
Gaskin 1982; Burger 1982; Nisbet 1983), gulls (e.g., Galusha and
Amlander 1978; Braune and Gaskin 1982; Burger 1982), ducks (Burger
et al. 1984), shorebirds (reviews in Burger 1984; Puttick 1984), and
herons (e.g., Bayer 1981). Terns may take advantage of different
stages of tide in different locations within an area, responding to
tidal effects on food availability (Braune and Gaskin 1982; Burger
1983), but falling and low tides often produce increased foraging
activity in terns (Becker et al. 1973; Nisbet 1983), gulls (Burger
1982), diving ducks (Burger et al. 1984), shorebirds (Burger 1984;
Puttick 1984), and herons (Bayer 1981). Ebbing and low water can
concentrate fish by making food available to them, such as in inlets
and river mouths, and by removing a large volume of the water that
enhances their cover at other times. This effect can be dramatic in
tide pools, such as occurred near Fire Island inlet. Ebbing and low
water also frequently stirs up the bottom, presumably increasing
the availability of invertebrates. Sand eels, at least, are most active
at the stage of tide providing the most food (Scott 1973).

Factors Affecting Formation of Tern Foraging Flocks

Fish Under Versus Adjacent to Terns

Terns foraged over patches of prey that were richer in food resources than adjacent waters, and this difference was highest at the time of the season when terns were raising chicks. This was true despite the fact that food was generally more plentiful during the incubation period (Safina and Burger ms. c). The energy requirements of breeding adults are much higher than those of nonbreeders (Furness 1982), and much of this energy demand occurs during chick provisioning when, for several weeks, they must commute to and from feeding areas, forage actively, and yield much of their catch to their young. If animals can more easily afford to violate maximizing decisions when food is plentiful (Sih 1982; Hodges 1985; Bailey 1986; and see above), then they should spend more effort maximizing returns when they are stressed. This may be part of the reason that the difference in food resources under versus adjacent to foraging flocks was greatest when terns were feeding chicks.

The contrast between prey under versus away from terns becomes weaker over the course of the season. This is partly a function of declining prey resources, but contributing to the effect is that terns have greater difficulty locating greater-than-ambient patches when patches are poor. Later in the season when prey is scarce, terns forage at patches containing prey at low densities that had been ignored earlier in the season, and foraged over prey at densities equal to, and in some instances below, densities in adjacent waters (Safina and Burger 1985).

Contrasts in prey density, abundance, and prey depth under versus away from terns were slightly weaker when bluefish were present than when they were absent. This may result from a tendency of bluefish to cause the fragmentation of prey fish schools and to drive prey to the surface (Safina and Burger ms. b). Thus, if schools are more uniformly spread out (broken up) and occupy a more uniformly shallow position in the water in response to bluefish presence, contrasts in concentration of prey under versus adjacent to tern flocks will be lower because prey are less concentrated. Alternatively, or perhaps relatedly, the lessening of the contrast between prey under versus away from terns in the presence of predatory fish may result from the fact that predatory fish presence generally coincided with lower prey numbers, and predators most often occurred later in the season, past the time of prey peaks when

contrasts were strongest. When less prey is present, there is necessarily a smaller difference between the amount of prey where prey
occurs versus where prey is absent, than there is when prey is
abundant.

During the terns' breeding season, prey contrasts (density and
abundance) under versus adjacent to feeding terns were greater when
bluefish were absent than when they were present. Postbreeding,
the contrast in prey fish density was still greatest when bluefish
were absent, but the contrast in prey abundance was greater when
bluefish were present. This suggests the increasing importance of
bluefish to terns in concentrating the prey fish within an area as the
season progresses and prey become scarcer, and in allowing terns to
assess where fish are.

Patch choice by terns depends in part on the inherent behavioral
tendencies of individual prey species. Examining the two principal
prey species (sand eels and anchovies) separately showed that sand
eel density and abundance were both greater under tern flocks than
adjacent to them, whereas anchovy density but not abundance was
greater under terns. Anchovies tended to remain deeper than sand
eels while in our study area (Safina and Burger ms. a). If deep schools
are detected by terns because of the few individuals that approach
the surface, and if the number of prey that approach the surface is
density dependent, then denser schools will be more detectable, and
offer more opportunities for prey capture, than larger but more
spread-out aggregations. Tern flocks often persisted over deep anchovy schools with little success. In many of these cases, water
visibility was relatively poor and it was apparent that, although the
main body of the school was 6–8 meters below the surface, their
presence was detected because a few individuals came much closer
to the surface.

The Formation of Flocks Relative to Ambient Food Availability
  Comparisons of flock and control transects yield insights into the
choices terns make and their ability to amplify the ambient conditions for food availability. There were no differences in temperature
or sea surface conditions between the two transect types, but control transects had higher clarity. Temperature can be a major influence on fish movements (e.g., Lund and Maltezos 1970) and fish
availability to seabirds (Straty and Haight 1979). Relatively sharp
sea surface temperature gradients are associated with pelagic seabird concentrations (e.g., Abrams and Underhill 1986). However, on
the relatively finer scale at which we worked, and given the con-

straints on the foraging range of birds who must carry food to breeding colonies in their bills, birds could probably do little to improve upon certain physical conditions. The physical ocean sets the broad limits within which birds may try to maximize their efficiency in utilizing habitat resources (Obst 1985). Other variables, like water clarity, can vary enough over relatively small areas of ocean to allow differences to be selected. Birds selected areas with lower clarity, a condition associated with fish staying closer to the surface. Indeed, fish were closer to the surface in flock than in control transects. One may assume that they select for the presence of fish and not the clarity of water per se, but it is easy to imagine that if clarity and water color differences are easily seen from the air by searching terns (and they probably are), it would not take long for terns to be conditioned to recognize those physical aspects of their environment which in their experience have a greater chance of yielding food, and to concentrate their efforts there before food is actually located. Such a strategy would help to reduce a vast volume of ocean into relatively few promising habitat patches, decreasing the area and the time required for sampling. For terrestrial animals, the use of learned food-indicating cues in selecting a habitat patch (such as a hawk landing in a tree alongside an old field) is such a basic fact that we hardly notice it as such.

Terns were able to locate prey and predatory fish concentrations that were much higher than those in control transects. Flock and control transects reflected a similar decline in food over the course of the season; however, the decline of prey was stronger in controls. Thus, tern foraging responses, while constrained by ambient physical and biotic patterns, nonetheless allow terns to locate, assess, and utilize areas in which food resources are of higher quality than in surrounding waters.

■

Summary

The distribution and dynamics of prey is of profound importance to the life cycle and evolution of oceanic seabirds. However, studies of prey dynamics and the trophic ecology of seabirds have lagged behind advances made in understanding seabird breeding biology, largely due to the difficulties of sampling prey (particularly highly mobile prey) in deep water, and because of the generally difficult logistics of working on the ocean. The lack of data on fish dynamics has repeatedly hampered the conclusions that could be drawn about the ways in which food affects seabird distribution, migration, social behavior, and reproduction.

Using sonar to obtain data on fish distribution, our studies were designed to investigate the dynamics of the fish community, factors that affect prey availability to common terns, and the ways in which terns respond to changing food availability. In the system in which we worked (in the ocean within 10 km of Fire Island Inlet, New York), several species of prey fish were important to terns. Their absolute and relative numbers changed within and among years. One predatory fish, the bluefish, which usually arrived in late spring, several weeks after the terns arrived, appeared to have a major effect on the prey community, causing sharp declines in prey numbers and a significant shift in mean prey depth toward the surface.

We found that physical features affect fish in several ways. Prey fish and predatory fish (bluefish) tended to stay away from the surface in clear or deep water. Prey fish also avoided staying close to the bottom, where predators lurked, preferring to remain at mid-depths. Prey fish congregated at the inlet on ebbing tide, and bluefish seemed to respond to the aggregation of prey by arriving late in the ebb. Moderate wind and choppy surface conditions seemed to disperse prey fish. Bluefish were more likely to appear near shore in moderately windy conditions, perhaps due to the effect of relatively strong onshore winds in pushing up warm surface water from the south. However, surface choppiness appeared to depress bluefish feeding, although it did not alter their mean depth. Cloud cover did not affect prey fish, but bluefish were more likely to move toward shore and feed as cloud cover increased.

Changes in fish availability altered tern foraging behavior. Flock sizes and the total number of foraging terns in the study area was correlated with prey fish abundance and density, especially within 3 meters of the surface. Tern density was inversely related to prey abundance, indicating that terns preferred to space themselves when prey was widely available, but was positively correlated to prey density and mean prey depth. Flock size and total number of foraging terns was not correlated with mean prey depth. The lack of a relationship of bird numbers to mean prey depth is probably due to the fact that deeper schools tend to contain more prey, and that the amount of prey that approaches the surface (i.e., becomes available to terns) is proportional to the amount of prey in the school. Dividing prey abundance by mean prey depth was a better predictor of flock size than was either abundance, density, or depth of prey alone. Tern flock sizes and flock densities were larger when bluefish were close to the surface and chasing prey fish, a situation that terns exploited frequently and may have relied upon as prey fish numbers declined over time as a result of bluefish presence. Bluefish numbers and activities did not affect the total number of birds foraging in the area.

Overall tern diving success (number of fish caught per unit time) was significantly related to prey fish density, but not to prey abundance in an area, and was best predicted by prey density divided by mean prey depth. Unlike the rate of successful dives, rates of unsuccessful and aborted dives are significantly correlated to prey abundance, suggesting that terns are less "careful" when food is plentiful. The percent of dives that were successful was not related to prey fish variables. Terns did achieve a higher percent of successful dives when foraging over bluefish, but because they dove less when foraging over bluefish, they did not capture more fish per unit time, on average, when bluefish were present. The presence of bluefish changed the relationship between tern diving success and prey density and abundance. When bluefish were present, tern success rates were significantly correlated to prey abundance but not prey density, the opposite of when bluefish were absent. The presence of bluefish also retarded the decline of prey abundance over time of day in transects through foraging terns.

The rate of agonistic interactions among terns was related to poor prey availability near the surface and to the amount of hovering they were doing. Terns hovered most when they were in dense groups close to the water while tracking sporadically surfacing bluefish that were driving prey to the surface.

Terns sometimes defended ephemeral feeding territories within foraging flocks. They usually did this when prey were present but not readily available and the fish capture rate was low (as when most prey were too deep and there were no predatory fish to drive them to the surface). Such territories were generally defended by vocal threats as individual terns flew in elliptical patterns while searching within flocks, and lasted only as long as the individual maintained its position within a flock. The dynamics of terns arriving and leaving, terns crossing territories, and the fact that usually only a minority of terns attempted to defend foraging spaces, made the visual effect of such behavior very subtle.

Some physical variables affected tern foraging. Birds dove less and aborted more dives as water clarity increased, and tern flocks formed where clarity was less than it was in our control area. Cloud cover and sea surface choppiness (up to the moderate conditions in which we worked) were not significantly related to tern foraging success. Wind speed was positively related to tern density, perhaps because it hampered the ability of individual terns to locate prey. Wind speed was negatively correlated to the percent of dives that were successful, though its negative correlation to the number of successful dives per unit time was not statistically significant. Dividing prey density and abundance by wind speed was a better predictor of the tern fishing success rate than were prey density, prey abundance, or wind speed singly. Successful dives decreased over the course of the morning, as did bird numbers. Bird correlations with tide parallel fish correlations with tide; birds were more active and caught more fish on ebbing water.

Fish were generally more plentiful under foraging terns than adjacent to tern flocks. This was especially true when bluefish were not present. The presence of bluefish generally corresponded with decreased differences in prey fish abundance and density in waters under versus adjacent to foraging terns. An exception to this was that in the postfledging period (subsequent to July 15), when prey numbers were low, the presence of bluefish corresponded with increased differences in prey abundance under versus adjacent to tern flocks. This suggests the increasing importance of bluefish to terns in making prey available and facilitating prey location by terns as the season progresses and prey become scarcer. Bluefish density was greater under terns than adjacent to tern flocks, but bluefish abundance, school size, and mean depth was not.

The formation of tern flocks was almost always dependent on the presence of one or more of the following conditions which caused prey to approach the surface within range of diving terns: a physical feature that attracted prey, such as an inlet (especially during ebbing tide), drift lines, or tide rips; shoals that forced prey closer to the surface; the presence of food (i.e., dense plankton swarms) near the surface; or the presence of foraging bluefish driving prey to the surface. Tern flocks formed where prey fish and predatory fish were relatively dense and relatively shallow, as compared to fish in ambient waters. Despite this, fish measures where terns foraged reflected the same seasonal trends as ambient waters. While the rate at which terns capture food is constrained by changes in prey population trends, prey behavior, and physical factors, terns' methods of foraging allow them to evaluate a variety of patches and choose those with relatively high prey availability.

Acknowledgments

Many persons contributed logistically, intellectually, financially, and with their enthusiasm and skepticism during the field work, analysis, and writing stages of our studies of terns and fish. We thank: Ken Able, Joanne Cardinali, John and Nancy DeBellas, Robin Densmore, Marilyn England, Michael Gochfeld, Rita Halbeisen, Chris Hoogendyk, William Kolodnicki, Charles F. Leck, John C. Ogden, Glenn Paulson, Carlo and Rose Safina, Valerie Schwaroch, Hathaway Scully, Kelly Smith, Alexander Sprunt IV, Robert Steidl, Richard H. Wagner, Peter Warny, and David Witting. We also thank The Natalie P. Webster Trust, the Moriches Bay and South Shore chapters of the National Audubon Society, and Evinrude, Inc. for practical and financial assistance. This work could not have been completed without the spare room at 54 Hollywood Avenue.

References

Abrams, R. W. and L. G. Underhill. 1986. Relationships of pelagic seabirds with the southern ocean environment assessed by correspondence analysis. *Auk* 103:221–225.

Anderson, D. W., F. Gress, and K. Mais. 1982. Brown Pelicans: influence of food supply on reproduction. *Oikos* 39:23–31.

Andersson, M., F. Gotmark, and C. G. Wiklund. 1981. Food information in the Black-headed gull, *Larus ridibundus*. *Behav. Ecol. Sociobiol.* 9:199–202.

Ashmole, N. 1963. The regulation of numbers of tropical oceanic birds. *Ibis* 103:458–473.

Ashmole, N. P. 1971. Seabird ecology and the marine environment. In D. S. Farner and J. R. King, eds., *Avian Biology*, 1:223–286. New York: Academic Press.

Ashmole, N. P. and M. J. Ashmole. 1967. Comparative feeding ecology of sea birds of a tropical oceanic island. *Peabody Mus. Nat. Hist., Yale Univ. Bull.* 24:1–131.

Atwood, J. L. and P. R. Kelly. 1984. Fish dropped on breeding colonies as indicators of Least Tern food habits. *Wils. Bull.* 96:34–47.

Au, D. W. K. and R. L. Pitman. 1986. Seabird interactions with dolphins and tuna in the eastern tropical Pacific. *Condor* 88:304–317.

Bailey, P. C. E. 1986. The feeding behaviour of a sit-and-wait predator, *Ranata dispar* (Heteroptera: Nepidae): optimal foraging and feeding dynamics. *Oecologia* 68:291–297.

Baltz, D. M., G. V. Morejohn, and B. S. Antrim. 1979. Size selective predation and food habits of two California terns. *Western Birds* 10:17–24.

Bayer, R. D. 1981. Arrival and departure frequencies of Great Blue Herons at two Oregon estuarine colonies. *Auk* 98:589–595.

Bayer, R. D. 1982. How important are bird colonies as information centers? *Auk* 99:31–40.

Becker, P. H., P. Finck, and A. Anlauf. 1985. Rainfall preceding egg-laying—a factor of breeding success in Common Terns *(Sterna hirundo)*. *Oecologia* 65:431–436.

Bennett, D. V. and F. A. Streams. 1986. Effects of vegetation on Notonecta (Hemiptera) distribution in ponds with and without fish. *Oikos* 46:62–69.

Black, B. B. and L. D. Harris. 1983. Feeding habitat of Black Skimmers wintering on the Florida gulf coast. *Wils. Bull.* 95:404–415.

Bourne, W. R. P. 1983. Birds, fish, and offal in the North Sea. *Mar. Poll. Bull.* 8:294–296.

Bovino, R. R. and E. H. Burtt, Jr. 1979. Weather-dependent foraging of Great Blue Herons *(Ardea herodias)*. *Auk* 96:628–630.

Braune, B. M. and D. E. Gaskin. 1982. Feeding ecology of nonbreeding populations of larids off Deer Island, New Brunswick. *Auk* 99:67–76.

Briggs, K. T., D. B. Lewis, W. B. Tyler, and G. L. Hunt, Jr. 1981. Brown Pelicans in southern California: habitat use and environmental fluctuations. *Condor* 83:1–15.

Brown, R. G. B. 1980. Seabirds as marine animals. In J. Burger, B. Olla, and H. E. Winn, eds., *Behavior of Marine Animals*, vol. 4: *Marine Birds*, pp. 1–40. New York: Plenum Press.

Buckley, F. G. and P. A. Buckley. 1974. Comparative feeding ecology of wintering adult and juvenile Royal Terns (Aves: Laridae, Sterninae). *Ecology* 55:1053–1063.

Buckley, F. G. and P. A. Buckley. 1980. Habitat selection and marine birds. In J. Burger, B. Olla, and H. E. Winn, eds., *Behavior of Marine Animals*, vol. 4: *Marine Birds*, pp. 69–112. New York: Plenum Press.

Burger, J. 1980. The transition to independence and postfledging parental care in seabirds. In J. Burger, B. Olla, and H. E. Winn, eds., *Behavior of Marine Animals*, vol. 4: *Marine Birds*, pp. 367–447. New York: Plenum Press.

Burger, J. 1981. A model for the evolution of mixed-species colonies of Ciconiiformes. *Quar. Rev. Biol.* 56:143–167.

Burger, J. 1982. Jamaica Bay studies, 1: Environmental determinants of abundance and distribution of Common Terns *(Sterna hirundo)* and Black Skimmers *(Rynchops niger)* at an east coast estuary. *Colonial Waterbirds* 5:148–160.

Burger, J. 1983. Jamaica Bay studies, 3: Abiotic determinants of distribution and abundance of gulls *(Larus)*. *Estuarine, Coastal, and Shelf Science* 16:191–216.

Burger, J. 1984. Abiotic factors affecting migrant shorebirds. In J. Burger, and B. Olla, eds., *Behavior of Marine Animals*, vol. 6: *Shorebirds: Migration and foraging behavior*, pp. 1–73. New York: Plenum Press.

Burger, J. and M. Gochfeld. 1983. Feeding behavior in Laughing Gulls: compensatory site selection by young. *Condor* 85:467–473.

Burger, J., J. R. Trout, W. Wander, and G. S. Ritter. 1984. Jamaica Bay studies, 7: Factors affecting the distribution and abundance of ducks in a New York estuary. *Estuar. Coast. Shelf Sci.* 19:673–689.

Caraco, T. and L. L. Wolf. 1975. Ecological determinants of group sizes of foraging lions. *Am. Nat.* 109:343–352.

Carpenter, F. L. and R. E. Macmillen. 1976. Threshold model of feeding territoriality and test with Hawaiian honeycreeper. *Science* 194:639–642.

Charnov, E. L., G. H. Orions, and K. Hyatt. 1976. Ecological implications of resource depression. *Am. Nat.* 110:247–259.

Clark, C. W. and M. Mangel. 1984. Foraging and flocking strategies: information in an uncertain environment. *Am. Nat.* 123:626–641.

Cody, M. L. 1974. Optimization in Ecology. *Science* 183:1156–1164.

Colblentz, B. E. 1985. Mutualism between Laughing Gulls *(Larus atricilla)* and epipelagic fishes. *Cormorant* 13:61–63.

Cosper, E. M., W. C. Dennison, E. J. Carpenter, V. M. Bricelj, J. G. Mitchell, S. H. Kuenstner, D. Colflesh, and M. Dewwy. 1987. Recurrent and persistent brown tide blooms perturb coastal marine ecosystem. *Estuaries* 10:284–290.

Courtney, P. A. and H. Blokpoel. 1980. Food indicators of food availability for Common Terns on the lower Great Lakes. *Can. J. Zool.* 58:1318–1323.

Cowan, P. J. 1972. The contrast and coloration of sea-birds: an experimental approach. *Ibis* 114:390–393.

Duffy, D. C. 1983. The foraging ecology of Peruvian seabirds. *Auk* 100:800–810.

Duffy, D. C. 1986. Foraging at patches: interactions between Common and Roseate Terns. *Ornis Scand.* 17:47–52.

Dunn, E. K. 1972. Effect of age on the fishing ability of Sandwich Terns. *Ibis* 114: 360–366.

Dunn, E. K. 1973. Changes in fishing ability of terns associated with windspeed and sea surface conditions. *Nature* 244:520–521.

Eadie, M. and A. Keast. 1982. Do goldeneye and perch compete for food? *Oecologia* 55:225–230.

Engen, S. and N. C. Stenseth. 1984. An ecological paradox: a food type may become more rare in the diet as a consequence of being more abundant. *Am. Nat.* 124:352–359.

Eriksson, M. O. G. 1979. Competition between freshwater fish and goldeneyes *Bucephala clangula* (L.) for common prey. *Oecologia* 41:99–107.

Eriksson, M. O. G. 1983. The role of fish in the selection of lakes by nonpiscivorous ducks: Mallard, Teal and Goldeneye. *Wildfowl* 34:27–32.

Eriksson, M. O. G. 1985. Prey detectability for fish-eating birds in relation to fish density and water transparency. *Ornis Scand.* 16:1–7.

Erwin, R. M. 1977. Foraging and breeding adaptations to different food regimes in three seabirds: the Common Tern, *Sterna hirundo*, Royal Tern, *Sterna maxima*, and Black Skimmer, *Rynchops niger*. *Ecology* 58:389–397.

Erwin, R. M. 1978. Coloniality in terns: the role of social feeding. *Condor* 80:211–215.

Erwin, R. M. 1983a. Feeding behavior and ecology of colonial waterbirds: a synthesis and concluding comments. *Colonial Waterbirds* 6:73–82.

Erwin, R. M. 1983b. Feeding habitats of nesting wading birds: spatial use and social influences. *Auk* 100:960–970.

Erwin, R. M. 1985. Foraging decisins, patch use and seasonality in egrets (Aves: Ciconiiformes). *Ecology* 66:837–844.

Evans, R. M. 1982. Foraging-flock recruitment at a Black-billed Gull colony: implications for the information center hypothesis. *Auk* 99:24–30.

French, T. W. 1981. Fish attack on Black Guillemot and Common Eider in Maine. *Wils. Bull.* 93:279–280.

Furness, R. W. 1982. Competition between fisheries and seabird communities. In J. H. S. Blaxter, F. S. Russell, and M. Yonge, eds. *Advances in Marine Biology*, 20:225–307. London: Academic Press.

Galusha, J. G., Jr. and C. J. Amlander, Jr. 1978. The effects of diurnal and tidal periodicities in the numbers and activities of Herring Gulls *Larus argentatus* in a colony. *Ibis* 120:322–328.

Glasser, J. W. 1984. Is conventional foraging theory optimal? *Am. Nat.* 124:900–905.

Gochfeld, M. 1978. Observations on feeding ecology and behavior of Common Terns. *Kingbird* 28:84–89.

Gochfeld, M. 1980. Mechanisms and adaptive value of reproductive synchrony in colonial seabirds. In J. Burger, B. Olla, and H. E. Winn, eds., *Behavior of Marine Animals*, vol. 4: *Marine Birds*, pp. 207–270. New York: Plenum Press.

Gochfeld, M. and J. Burger. 1982. Feeding enhancement by social attraction in the Sandwich Tern. *Behav. Ecol. Sociobiol.* 10:15–17.

Gould, P. J. 1974. Sooty Tern *(Sterna Fuscata)*. *Smithsonian Contributions to zoology* 158:5–52.

Hamilton, W. D. 1971. Geometry for the selfish herd. *J. Theor. Biol.* 31:295–311.

Hafner, H. and R. H. Britton. 1983. Changes of foraging sites by nesting Little Egrets (Egretta garzetta L.) in relation to food supply. *Colonial Waterbirds* 5:24–30.

Harris, M. P. 1965. The food of some Larus gulls. *Ibis* 107:43–53.

Harrison, C. S., T. S. Hida, and M. P. Seki. 1983. Hawaiian seabird feeding ecology. *Ecological Monographs* 85:1–71.

Heinrich, B. 1983. Do Bumblebees forage optimally, and does it matter? *Am. Zool.* 23:273–281.

Hewett, R. P., P. E. Smith, and J. C. Brown. 1976. Development and use of sonar mapping for pelagic stock assessment in the California current area. *U.S. Fishery Bull.* 74:281–297.

Hodges, C. M. 1985. Bumblebee foraging: the threshold departure rule. *Ecology* 66:179–187.

Hoffman, W., D. Heinemann, and J. A. Wiens. 1981. The ecology of seabird feeding flocks in Alaska. *Auk* 98:437–456.

Hopkins, C. D. and R. H. Wiley. 1972. Food parasitism and competition in two terns. *Auk* 89:583–594.

Horn, H. S. 1968. The adaptive significance of colonial nesting in the Brewer's blackbird *(Euphagus cyanocephalus)*. *Ecology* 49:682–694.

Hunt, G. L. and M. W. Hunt. 1976. Gull chick survival: the significance of growth rates, timing of breeding and territory size. *Ecology* 57:62–75.

Hunter, M. L., Jr., J. J. Jones, K. E. Gibbs, and J. Moring. 1986. Duckling responses to lake acidification: do black ducks and fish compete? *Oikos* 47:26–32.

Janetos A. C. and B. J. Cole. 1981. Imperfectly optimal animals. *Behav. Ecol. Sociobiol.* 9:203–209.

Janzen, D. 1986. Science is forever. *Oikos* 46:218–283.

Koslow, J. A. 1981. Feeding selectivity of schools of northern anchovy in the southern California bight. *U.S. Fishery Bull.* 79:131–142.

Krebs, J. R., A. I. Houston, and E. L. Charnov. 1981. Some recent developments in optimal foraging. In A. C. Kamil and T. D. Sargent, eds., *Foraging Behavior*, pp. 3–18. New York: Garland Publishing.

Krebs, J. R., D. W. Stephens, and W. J. Sutherland. 1983. Perspectives in optimal foraging. In A. H. Brush and G. A. Clark, Jr., *Perspectives in Ornithology*, pp. 165–216. Cambridge: Cambridge University Press.

Kress, S. T., E. H. Weinstein, and I. C. T. Nisbet. 1983. The status of tern populations in northeastern United States and adjacent Canada. *Colonial Waterbirds* 6:84–106.

Kuhlmann, D. H. H. and H. Karst. 1967. Open-water observations on the behaviour of sand eel schools in the Western Baltic. *Z. Tierpsychol.* 24:282–297.

Kushlan, J. A. 1976. Wading bird predation in a seasonally fluctuating pond. *Auk* 93:464–476.

Kushlan, J. A. 1977. The significance of plumage color in the formation of feeding aggregations of Ciconiiformes. *Ibis* 119:361–364.

Kushlan, J. A. 1978. Feeding ecology of wading birds. In A. Sprunt, J. C. Ogden, and S. Winckler, eds., *Wading Birds*, pp. 249–298. New York: National Audubon Society.

Kushlan, J. A. 1979. Feeding ecology and prey selection in the White Ibis. *Condor* 81:376–389.

Kushlan, J. A. 1981. Resource use strategies of wading birds. *Wils. Bull.* 93:145–163.

Lack, D. 1968. *Ecological Adaptations for Breeding in Birds.* London: Methuen.

Laskey, A. R. 1935. Mockingbird life history studies. *Auk* 52:370–381.

Lederer, R. J. 1977. Winter feeding territories in the Townsend's solitaire. *Bird-Banding* 48:11–18.

Levinton, J. S. 1982. *Marine Ecology.* Englewood Cliffs, N.J.: Prentice-Hall.

Lund, W. A., Jr., and G. C. Maltezos. 1970. Movements and migrations of the Bluefish, *Pomatomus saltatrix*, tagged in waters of New York and southern New England. *Trans. Am. Fish. Soc.* 99:719–725.

Lyman, H. and F. Woolner. 1954. *Striped Bass Fishing.* New York: Barnes.

Madsen, K. P. 1963. Tides and the sand eel fishery. *Skr. Dan. Fisk. Havunders.* 23:45–48.

Michener, J. R. 1951. Territorial behavior and age composition in a population of Mockingbirds at a feeding station. *Condor* 53:276–283.

Milinski, M. 1984. A predator's costs of overcoming the confusion-effect of swarming prey. *Anim. Behav.* 32:1157–1162.

Miller, N. E. 1941. The frustration-aggression hypothesis. *Psychol. Rev.* 48:337–342.

Miller, L. and J. Confer. 1982. A study of the feeding of young Common Terns at one site in Oneida Lake during 1980. *Kingbird* 32:167–172.

Munger, J. C. 1984. Optimal foraging? Patch use by horned lizards *(Iguanidae: Phrynosoma). Am. Nat.* 123:654–680.

Murphy, E. C., R. H. Day, K. L. Oakley, and A. A. Hoover. 1984. Dietary changes and poor reproductive performance in Glaucous-winged Gulls. *Auk* 101:532–541.

Meyers, J. P. 1983. Commentary. In A. H. Brush and G. A. Clark, Jr., eds., *Perspectives in Ornithology*, pp 216–221. Cambridge: Cambridge University Press.

Nelson, J. B. 1970. The relationship between behavior and ecology in the Sulidae. *Oceanog. Mar. Biol. Ann. Rev.* 8:501–574.

Nisbet, I. C. T. 1973. Courtship-feeding, egg-size, and breeding success in Common Terns. *Nature* 241:141–142.

Nisbet, I. C. T. 1977. Courtship-feeding and clutch size in Common Terns, *Sterna hirundo*. In B. Stonehouse and C. Perrins, eds., *Evolutionary Ecology*, pp. 101–109. Baltimore, Md.: University Park Press.

Nisbet, I. C. T. 1978. Dependence of fledging success on egg-size, parental performance and egg-composition among Common and Roseate Terns, *Sterna hirundo* and *S. dougallii. Ibis* 120:207–215.

Nisbet, I. C. T. 1981. Biological characteristics of the Roseate Tern, *Sterna dougallii.* Washington, D.C.: U.S. Department of the Interior, Fish and Wildlife Service, Office of Endangered Species.

Nisbet, I. C. T. 1983. Territorial feeding by Common Terns. *Colonial Waterbirds* 6:64–70.

Nisbet, I. C. T. and M. E. Cohen. 1975. Asynchronous hatching in Common and Roseate Terns, *Sterna hirundo* and *S. dougallii. Ibis* 117:374–379.

Obst, B. S. 1985. Densities of Antarctic seabirds at sea and the presence of the krill *Eupausia superba. Auk* 102:540–549.

Ogden, J. C. and S. A. Nesbit. 1979. Recent Wood Stork population trends in the United States. *Wils. Bull.* 91:512–523.

Ogden, J. C., J. A. Kushlan, and J. T. Tilmant. 1976. Prey selectivity by the Wood Stork. *Condor* 78:324–330.

Parkin, D. T., A. W. Ewing, and H. A. Ford. 1970. Group diving in the Blue-footed Booby. *Ibis* 112:111–112.

Parsons, J. 1976. Factors determining the number and size of eggs laid by the Herring Gull. *Condor* 78:481–492.

Pearson, T. H. 1968. The feeding biology of sea-bird species breeding on the Farne Islands, Northumberland. *J. Animal Ecol.* 37:521–551.

Pehrsson, O. 1984. Relationships of food to spatial and temporal breeding strategies of mallards in Sweden. *J. Wild. Manage.* 48:322–339.

Perrins, C. M. 1970. The timing of birds' breeding seasons. *Ibis* 112:242–255.

Pitcher, T. J. and B. L. Partridge. 1979. Fish school density and volume. *Marine Biol.* 54:383–394.

Pitcher, T. J. and C. J. Wyche. 1983. Predator-avoidance behavior of sand-eel schools: why schools seldom split. In L. G. Noakes et al., eds, *Predators and Prey in fishes*, pp. 193–204. The Hague: Junk.

Porter, J. M. and S. G. Sealy. 1982. Dynamics of multispecies feeding flocks: age-related feeding behavior. *Behaviour* 81:91–109.

Powell, G. V. N. 1983. Food availability and reproduction by Great White Herons, *Ardea herodias*: a food addition study. *Colonial Waterbirds* 6:139–147.

Powell, G. V. N. 1985. Sociobiology and adaptive significance of interspecific foraging flocks in the neotropics. In P. A. Buckley et al., eds., *Neotropical Ornithology*, pp. 713–732. *Ornithol. Managr.* 36.

Power, M. E. 1984. Depth distributions of armored catfish: predator-induced resource avoidance? *Ecology* 65:523–528.

Pratt, H. M. 1980. Direction and timing of Great Blue Heron foraging flights from a California colony: implications for social facilitation of food finding. *Wils. Bull.* 92:489–496.

Puttick, G. M. 1984. Foraging activity patterns in wintering shorebirds. In J. Burger, and B. Olla, eds., *Behavior of Marine Animals*, vol. 6: *Shorebirds: Migration and Foraging Behavior*, pp. 203–232. New York: Plenum Press.

Raynor, G. S. 1972. Overland feeding flights by the Common Tern on Long Island. *Kingbird* 22: 63–71.

Richards, S. W. 1976. Age, growth, and food of Bluefish *(Pomatomus salatrix)* from east-central Long Island Sound from July through November 1975. *Trans. Am. Fish. Soc.* 105:523–525.

Romero, A. 1985. Cave colonization by fish: role of bat predation. *Am. Midl. Nat.* 113:7–17.

Ryder, J. P. 1980. The influence of age on the breeding biology of colonial seabirds. In B. Olla, J. Burger, and H. E. Winn, eds., *Behavior of Marine Animals*, vol. 4: *Marine Birds*, pp. 153–156. New York: Plenum Press.

Safina, C. and J. Burger. 1985. Common tern foraging: seasonal trends in prey fish densities and competition with bluefish. *Ecology* 66:1457–1463.

Safina, C. and J. Burger. Manuscript a. Habitat partitioning by sandeels and bay anchovies: alternate predator avoidance strategies?

Safina, C. and J. Burger. Manuscript b. Population interactions among free-living bluefish and prey fish in an ocean environment.

Safina, C. and J. Burger. Manuscript c. Prey dynamics and the breeding phenology of Common Terns: constraints on adaptive evolutionary response?

Safina, C. and J. Burger. 1988. Use of sonar and a small boat for studying foraging ecology of seabirds. *Colonial Waterbirds*.

Salt, G. W. and D. E. Willard. 1971. The hunting behavior and success of Forster's Tern. *Ecology* 52:989–998.

Schaffner, F. C. 1986. Trends in Elegant Tern and Northern Anchovy populations in California. *Condor* 88:347–354.

Schluter, D. 1982. Seed and patch selection by galapagos ground finches: relation to foraging efficiency and food supply. *Ecology* 63:1106–1120.

Schluter, D. 1984. Feeding correlates of breeding and social organization in two Galapagos finches. *Auk* 101:59–68.

Schreiber, R. W. and E. A. Schreiber. 1984. Central Pacific seabirds and the El Niño southern oscillation: 1982 to 1983 perspectives. *Science* 225:713–716.

Scott, D. 1984. The feeding success of cattle egrets in flocks. *Anim. Behav.* 32:1089–1100.

Scott, J. S. 1973. Food and inferred feeding behavior of northern sand lance (Ammodytes dubius). *J. Fish. Res. Bd. Can.* 30:451–454.

Sealy, S. G. 1973. Interspecific feeding assemblages of marine birds off British Columbia. *Auk* 90:796–802.

Sibly, R. M. 1984. Optimal group size is unstable. *Anim. Beh.* 31:947–948.

Sih, A. 1980. Optimal behavior: can foragers balance two conflicting demands? *Science* 210:1041–1043.

Sih, A. 1982. Optimal patch use: variation in selective pressure for efficient foraging. *Am. Nat.* 120:666–685.

Sih, A. 1984. The behavioral response race between predators and prey. *Am. Nat.* 123:143–150.

Simon, C. A. 1975. The influence of food abundance on territory size in the iguanid lizard *Sceloporus jarrovi*. *Ecology* 56:993–998.

Stenger, J. 1958. Food habits and available food of ovenbirds in relation to territory size. *Auk* 75:335–346.

Straty, R. R. and R. E. Haight. 1979. Interactions among marine birds and commercial fish in the eastern Bering Sea. In J. C. Bartonek and D. N. Nettleship, eds., *Conservation of Marine Birds of Northern North America*, pp. 201–219. Wash. D.C.: U.S. Department of the Interior, Fish and Wildlife Service, Research Report #11.

Tinbergen, N. 1952. "Derived" activities, their causation, biological significance, and emancipation during evolution. *Quart. Rev. Biol.* 27:1–32.

Veen, J. 1977. Functional and causal aspects of nest distribution in colonies of the Sandwich Tern. *Beh. Suppl.* 20:1–193.

Ward, P. and A. Zahavi. 1973. The importance of certain assemblages of birds as "information-centres" for food-finding. *Ibis* 115:517–534.

Whoriskey, F. G. and G. L. Fitzgerald. 1985. The effects of bird predation on an estuarine stickleback (Pisces: Gasterosteidai) community. *Can. J. Zool.* 63:301–307.

Wilson, E. O. 1975. *Sociobiology*. Cambridge: Bellknap/Harvard.

Woodby, D. A. 1984. The April distribution of murres and prey patches in the southeastern Bering Sea. *Limnol. Oceanogr.* 29:181–188.

Zach, R. and J. B. Falls. 1976. Ovenbird (Aves: parulidae) hunting behavior in a patchy environment: an experimental study. *Can J. Zool.* 54:1863–1879.

Zach, R. and J. M. Smith. 1981. Optimal foraging in wild birds? In A. C. Kamil and T. D. Sargent, eds., *Foraging Behavior*, pp. 95–109. New York: Garland.

Zaret, T. M. and R. T. Paine. 1973. Species introduction in a tropical lake. *Science* 182:449–455.

5

■ Seabird Relationships with Tropical Tunas and Dolphins

David W. Au and Robert L. Pitman • *National Marine Fisheries Service, Southwest Fisheries Center, La Jolla, California*

Flocks of seabirds accompanying surface-schooling tunas are characteristic of tropical seas, but are especially notable in the eastern tropical Pacific (ETP) where the birds also associate with dolphins. We recently described these species interactions (Au and Pitman 1986) from extensive ship surveys that entailed the examination of cetacean schools, particularly those associated with birds and tuna. Our observations enabled us to form an ecological perspective of species behaviors and interrelationships. There have been few other studies on the oceanic ecology of ETP seabirds, although the birds are much watched by fishermen as indicators of fish and fishing conditions. King (1970, 1974a) and Gould (1971) presented the first comprehensive descriptions of tropical central and eastern Pacific birds at sea, but as their observations were primarily from ships conducting oceanographic surveys, they had limited opportunities for closeup observations of feeding flocks and were not able to study the relationships with cetaceans.

In this essay we review the relationships between seabirds, tuna, and cetaceans as observed in the ETP and describe the ecological role of tunas. We will explain how the organization of, and interactions within, the apex pelagic community might depend upon forage and foraging tactics, especially that of the tunas—perhaps the key top predators of tropical seas. Our inferences will be based largely upon the behavior of the birds and dolphins, as the tuna were seldom directly observable.

■
Data and Methods

Observations on birds, tuna, and cetaceans were obtained from both biological census and oceanographic ship surveys. Most important were the Southwest Fisheries Center (SWFC) dolphin surveys of 1976, 1977, 1979, and 1980, designed to assess the distributions and abundances of dolphins involved in the "porpoise-tuna" fishery of the eastern Pacific. These surveys provided information over a broad area overlapping the tuna fishing grounds. In addition to our participation on these biological surveys, Pitman studied birds, tuna, and cetaceans from ships conducting physical oceanographic studies in the central and eastern Pacific.

The search and much of the observations were conducted through twenty or twenty-five power binoculars, generally mounted both port and starboard on or above the flying bridge of each survey ship. We usually searched between 6 A.M. and 6 P.M., using two teams of observers. The high powered binoculars proved indispensable for closeup observations and for minimizing the overlooking of bird flocks and cetacean schools. On the dolphin surveys, most mammal schools were approached closely after initial detection for better observations (often within a hundred meters), as were bird flocks if they appeared to be associated with cetaceans. Flocks and schools were usually not approached on the oceanographic surveys. On all surveys, species were identified whenever possible, and numbers estimated for all flocks and mammal schools. We took notes on any tuna seen, but direct observations or measurements of these fish schools were not feasible.

Noon positions from the above cruises, shown in figure 5.1, illustrate survey coverage. The concentrations of survey days along certain lines is due to repeated hydrographic transects on the oceanographic cruises. Though there were surveys during every month, about 63 percent of the observations took place during January through March; there is thus a seasonal bias in our data. A monthly breakdown of observations is given by Pitman (1986), and a more detailed description of the dolphin surveys by Au and Perryman (1985).

■
Flock Characteristics

We defined a flock as an aggregate of ten or more birds. We did not include storm-petrels, phalaropes, tropic birds, or the occasional gulls, as they feed largely independently of fish schools in pelagic waters and rarely occur with either tuna or cetaceans. The 1977, 1979, and 1980 dolphin surveys provided the most representative and accurate subset of our observations; these data are summarized in table 5.1 (flock associations) and table 5.2 (flock composition) by 5° latitude intervals and according to eastern and western sectors. We sighted a total of 637 flocks in the eastern sector, and 125 flocks in the western sector (table 5.1). Overall, few (25 percent) of these flocks were with cetaceans, although the association rates were high in certain areas (see below). If we assume that flocks of ten or more birds are associated with tuna, most tuna schools, then, were not with cetaceans.

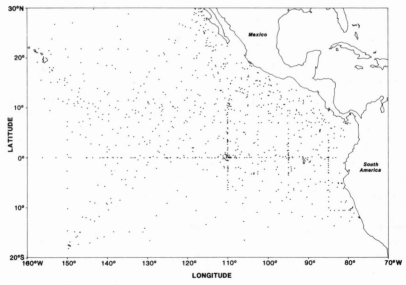

Figure 5.1. Noon positions of sea days during which a seabird watch was maintained.

Table 5.1. Percentage of seabird flocks associated with whale or delphinid cetaceans by area.

Center of 5° Latitude Intervals	EASTERN SECTOR (COAST–125°W)		Pct. flocks with		WESTERN SECTOR (> 125°W–155°W)		Pct. flocks with	
	Days observed	No. flocks	whales	delphinids	Days observed	No. flocks	whales	delphinids
25°N	16	6	0	0	21	7	0	0
20°N	14	11	0	73	15	19	0	0
15°N	35	104	0	68	27	23	0	0
10°N	54	133	2	35	22	51	0	30
5°N	49	126	2	27	11	13	0	8
0°	38	108	0	4	7	6	0	0
5°S	14	80	1	5	8	5	0	0
10°S	11	69	0	7	5	1	0	0
15°S			0	0			0	0
Total	231	637	1	27	116	125	0	9

NOTES: Flocks are defined as ≥ 10 birds; data from January–February cruises, 1977, 1979, 1980; percentages have been rounded; each latitude interval is defined by its center latitude ± 2.5°.

Table 5.2. Species composition and species diversity of flocks by area.

Sector	Center of 5° Latitude Intervals	Total Birds	PERCENT SPECIES COMPOSITION									SPECIES DIVERSITY	
			boobies	sooty tern	WTSW	jaeger	frigate birds	LPT	other terns	other shear-waters	other petrels	$\frac{1}{\Sigma Pi^2}$[a]	Spp. per Flock[b]
Eastern (From Coast to 125°W)	25°N	151	15	3	8	1	10			68	21	1.96	—
	20°N	1,279	50	4	30	11	+	2	22	46	3	3.46	4.4
	15°N	13,035	14	28	32	13	2	+	1	1	+	2.80	4.0
	10°N	6,852	42	20	13	2	8	1	1	2	+	4.40	4.2
	5°N	9,358	7	90	1	+	5	+	17	+	+	3.75	3.3
	0°	4,092	+	94		+	+	+	1	+		1.22	1.2
	5°S	3,696				+	2		4		+	1.12	1.3
	10°S	5,366	6	47		1	1	2	43		+	2.43	—
	15°S												

Western
(From > 125°W
to 155°W)

Latitude											Simpson's Index[a]	
25°N												
20°N	100	2	76	16	6	9	13	1	3	1	1.66	—
15°N	1,076	+	64	8	2		10	+	1	+	2.23	3.0
10°N	3,355	+	69	7	1		19	+	1	7	1.92	2.7
5°N	2,439	+	68	4	+		15	+	1	2	2.02	2.3
0°	4,239		88	4	+		4	1	1	1	1.29	2.2
5°S	6,983	1	97	1		1				1	1.07	
10°S	2,576	+	52	2	+	2		40	+	4	2.29	—
15°S	1,538	+	40	1		1		57	+	2	2.07	—

NOTES: Flocks are defined as ≥ 10 birds; flocks were evaluated whether or not cetaceans were present; data from January–February cruises, 1977, 1979, 1980; percentages have been rounded: each latitude interval is defined by its center latitude ± 2.5°.

Abbreviations: WTSW = wedge-tailed shearwater; LPT = large pterodroma; + = small percent < 1.0.

[a] Simpson's Index.

[b] Mean spp. per flock weighted by flock size; dash indicates unreliable data; for dolphin-associated flocks, mean spp./flock was 4.4 (s = 2.0) in eastern sector and 1.9 (s = 1.2) in western sector (Au and Pitman 1986).

Dichotomy in the Pelagic Community

Two distinct faunal communities involving seabird flocks occur
in the ETP, and suggest a strong, regional change either in prey or
in the responses of predators to prey. The dichotomy separates
multispecies flocks often associated with delphinid cetaceans from
virtually single-species flocks seldom with delphinids (i.e., dolphins
and other small, toothed "whales"). This separation occurs between
latitudes 0° and 5°N and approximately corresponds with the oceanographic division between the permanently warm (> 25°C) and low
salinity (< 34°/p.p.t.) Tropical (Surface) Water and the Equatorial
(> 34°/p.p.t.) and Southern Subtropical (> 35°/p.p.t.) waters (see
Wyrtki 1966; Ashmole 1971).

Northern (Tropical Water) Flocks
   The Northern, or Tropical Water, flocks in the eastern sector
were notable for frequently being with delphinid cetaceans: 68 percent and 73 percent of flocks from the 15°N and 20°N (± 2.5°)
latitude intervals, respectively, were with delphinids (table 5.1).
Associations with whales in these latitudes were infrequent (0–2
percent). These flocks (table 5.2) were typically multispecies aggre-
gates of boobies—primarily red-footed *(Sula sula)*, masked *(S. dac-
tylatra)*, and brown *(S. leucogaster)*—and wedge-tailed shearwaters
(Puffinus pacificus), sooty terns *(Sterna fuscata)*, and jaegers *(Stercorarius* spp.) (see appendix 5.1 for a list of species names). Boobies
were most abundant, composing 42 percent and 50 percent of individuals in flocks in the latitude intervals centered at 5°N and 15°N.
Their reduced importance about the 10°N interval was due in part
to sampling near the Costa Rica Dome, a localized upwelling regime (Wyrtki 1964) that does not normally produce good catches of
yellowfin tuna (the significance of which will be explained below).
   In the western sector the northern flocks were associated with
delphinids mainly within the 10°N interval (a much narrower zon-
ation than seen to the east) where the association rate was 30
percent. As in the east, whales appeared unimportant to seabirds;
none were seen with flocks. The species composition of these flocks
was different from that of flocks farther to the east, and thus com-
prised a second type of multispecies community consisting mainly
of sooty terns and Juan Fernandez/white-necked petrels *(Pterodroma externa externa/cervicalis*—subspecies that are difficult to
separate in the field).

Southern (Equatorial and Southern Subtropical Water) Flocks

South of the 5°N latitude interval, in Equatorial and Southern Subtropical waters, but primarily the latter, flocks in the eastern sector occurred infrequently with delphinid cetaceans (table 5.1), in sharp contrast to the northern multispecies flocks. The southern flocks were clearly dominated by sooty terns (table 5.2) which composed 90 percent or more of birds in flocks in the 0° and 5°S latitude intervals, mainly in the areas away from islands.

In the west none of the southern flocks encountered were associated with either delphinids or whales. The sooty tern was still the most abundant species, up to 97 percent of the birds in all flocks. White terns *(Gygis alba)* and noddy terns *(Anous* spp.) were an increasing component of the flocks in the far southern latitudes; the latter terns were especially abundant near islands.

Species Diversity Differences

The transition and difference between the northern and southern seabird communities are reflected in latitudinal changes in species diversity. Values of Simpson's Dominance Index $\frac{1/\Sigma_{Pi^2}}{Pi^2}$ where p_1 is the fraction of total birds that are species i) and of the average number of species per flock are given in the last two columns of table 5.2. Here the index could vary between 1.0 for flocks completely dominated by one species to 9.0 for flocks with individuals evenly divided among all nine categories of species. Simpson's Index is one of the more useful and easily understood measures for describing species in communities (Hill 1973) and shows that flocks in the northern latitude intervals, 20°N to 5°N, were on average 3.1 to 1.7 times more diverse than were flocks in the intervals 0° and 5°S, in the eastern and western sectors respectively. The southern flocks had index values close to 1.0 due to dominance by sooty terns. Data in the species per flock column suggest a decline in the average number of species from northern to southern latitudes (from ca. 4 to 1 spp./flock), at least in the eastern sector. Although there were difficulties in observing and identifying species from non–dolphin-associated flocks, this measure indicates areawide changes in diversity similar to those shown by Simpson's Index.

■
Seabirds and Cetaceans

As a group, seabirds associate only with specific cetaceans, a fact reflecting an underlying, probably feeding, relationship. This is shown in table 5.3, a summary of schooling and bird association characteristics of ETP whales and delphinids, again mainly from the more representative 1977, 1979, and 1980 dolphin surveys. Study of this table will give the reader an appreciation of the structure of the pelagic community under discussion, including the relative abundance of the different Cetacea and the likelihood (percent occurrence) of finding particular bird species in the associated flocks. In judging relative abundance, however, one should remember that different species vary in their detectability according to their size (school and individual) and behavior (including association with birds).

Associations with Dolphins

Of all Cetacea, spotted *(Stenella attenuata)* and spinner *(S. longirostris)* dolphins are associated most often with birds: 74 percent and 78 percent, respectively, of these species' schools were with flocks (table 5.3: first two species, col. 6). A high percentage of these schools included other cetaceans, 79 percent in the case of the spinner dolphin (col. 5). Most of this mixing involved these two species themselves: of the 113 spinner dolphin schools, 67 percent were mixed with spotted dolphins and of 206 spotted dolphin schools, 37 percent were mixed with spinner dolphins. Both of these dolphins occurred in large schools, averaging 150 and 133 individuals respectively (col. 3), and were associated with large flocks averaging 121 and 147 birds respectively (col. 7). The standard deviations (s) of these measures were large relative to the means, due to size distributions skewed toward the larger sizes. The bird species most likely present (high percent occurrence) in flocks with these dolphins were boobies *(Sula* spp.), frigate birds *(Fregata* spp.), wedge-tailed shearwaters, and jaegers, generally in that order. Spotted and spinner dolphins are diurnally active—i.e., fast swimming ("porpoising"), leaping often—and are frequently associated with yellowfin tuna. Both are pursued by purse seiners in "porpoise-tuna" fishing, a technique in which tuna are first caught with dolphins, then retained as the mammals are subsequently released.

Table 5.3. Summary of cetacean species encountered and their associations with seabirds: Characteristics of schools and flocks.

SPECIES	SCHOOLS					ASSOCIATED FLOCKS			COMMENTS
	n	$\bar{x}$	s	Pct. mixed[a]	Pct. with flocks	$\bar{x}$	s	Spp. of >50% occur[b]	
Spotted dolphin (Stenella attenuata)	206	150	177	40	74	121	219	booby −82% frigate −70 WTSW−57 jaeger −53	Main target of "porpoise-tuna" fishing; diurnally active (leaping, fast swimming) sp.; mixed schools usually with spinner dolphin. Frequently with flocks; epipelagic prey (Perrin et al. 1973).
Spinner dolphin (Stenella longirostris)	113	133	207	79	78	147	258	booby −73% frigate −69 jaeger −62 WTSW−54	Second most important dolphin in "porpoise-tuna" fishing; diurnally active, usually mixed with spotted dolphins; frequently with flocks. Epi- and mesopelagic prey (Perrin et al. 1973).
Common dolphin (Delphinus delphis)	62	129	176	6	39	38	46	booby −70%	Less involvement in "porpoise-tuna" fishing; often escapes with tuna from nets; very active diurnally; third most frequent dolphin with flocks. May feed on deep prey (Fitch and Brownell 1968).

Table 5.3. (continued)

SPECIES	SCHOOLS					ASSOCIATED FLOCKS			COMMENTS
	n	$\bar{x}$	s	Pct. mixed[a]	Pct. with flocks	$\bar{x}$	s	Spp. of > 50% occur[b]	
Striped dolphin (*Stenella coeruleoalba*)	139	54	62	9	7	15	24		Common, diurnally very active species in relatively small schools; seldom with birds or tuna. May feed on mesopelagic prey (Miyazaki et al. 1973).
Bottlenose dolphin (*Tursiops truncatus*)	119	25	59	41	13	11	12		Often with *Globicephala*, *Grampus*, *Steno*; sometimes with *Stenella*; often approaches ships; not with large flocks.
Risso's dolphin (*Grampus griseus*)	106	11	18	20	3	16	17		Often with *Tursiops*, *Globicephala*, traveling slowly. Seldom with birds.
Pilot whale (*Globicephala macrorhynchus*)	98	14	12	37	4	11	11		Often with *Tursiops*, *Grampus*; appears to be resting much of the time. Seldom with birds.
Rough-toothed dolphin (*Steno bredanensis*)	54	12	10	15	35	12	15	booby –72% frigate –72	Sluggish behavior; often near flotsam; sometimes with *Tursiops*, *Globicephala*, *Grampus*; fourth most frequent dolphin with birds.

Species	n	x̄	s					Behavior
Melon-headed whale (Peponocephala electra)	4 [42	194	151	75 17	25 7]c	1	—	Diurnally active; often in large, dense schools with *Lagenodelphis*. Seldom with birds.
Fraser's dolphin (Lagenodelphis hosei)	10 [23	391	576	40 44	20 13]c	2	—	Often in large, densely packed, fast-moving schools with *Peponocephala*. Schools often execute coordinated turning and diving.
Pygmy killer whale (Feresa attenuata)	9	37	18	0	0	—	—	Appears to rest during the day; may approach ships. Seldom with birds.
False killer whale (Pseudorca crassidens)	19	13	13	26	10	12	5	Sometimes with *Tursiops*; will approach ships at high speeds; not with large flocks.
Killer whale (Orcinus orca)	8	7	4	12	12	20	—	Small schools, usually loosely aggregated; not with large flocks.
Rorquals (baleen whales)[d]	94	2	1	3	3	43	27	Generally feed near surface on small fish and crustaceans; seldom with birds.
Ziphiids (beaked whales)	85	2	2	1	0	—	—	Deep feeder; shy and inconspicuous. Not with birds.
Sperm whale (Physeter macrocephalus)	63	8	7	8	2	2	—	Deep feeders; sometimes in large aggregations. Rarely with birds.
Dwarf/pygmy sperm whale (Kogia spp.)	30	2	1	0	0	—	—	Difficult to see except in calm seas. Not with birds.

NOTES: Numbers in table rounded for greater clarity; standard deviations not calculated for n < 10; birds counted regardless of flock size; n = number of schools, x̄ and s refer to school or flock size. Abbreviations, *see* table 5.2.

[a] i.e., school mixed with other cetacean sp.

[b] Listed here are bird spp. and their percentage occurrence in flocks associated with the given cetacean sp., provided the percentage was ≥ 50 and flocks ≥ 10 birds. WTSW = wedge-tailed shearwater.

[c] From larger sample from all research ship observations, and also including *Feresa* schools.

[d] Mostly Bryde's whale.

The common dolphin *(Delphinus delphis)* was the third most frequent species of dolphin found with bird flocks, although the percentage of its schools found with flocks was much lower (39 percent). This species also forms large ($\bar{x}$ = 129), actively porpoising schools. Boobies, and also wedge-tailed shearwaters and frigate birds, often occurred in flocks associated with this dolphin. Fishermen do not regularly catch tuna with the common dolphin.

The last of the more frequently encountered, active delphinids, the striped dolphin *(Stenella coeruleoalba)*, is seldom found with birds. It usually occurred in fast-moving and often high-leaping, relatively small, unmixed schools ($\bar{x}$ = 54) that were without indications of associated tuna. It is seldom deliberately fished on by purse seiners.

With two exceptions (see below) the remaining delphinids occur in small schools that are seldom with birds, and perhaps never with large flocks. Except for the false killer whale *(Pseudorca crassidens)*, all are relatively slow moving. And although the sluggish-behaving rough-toothed dolphin *(Steno bredanensis)* would seem to be an exception in that 35 percent of its schools were recorded with birds, we saw no evidence to indicate that this or any other other of these remaining dolphins occurred with tuna.

The melon-headed whale *(Peponocephala electra)*—sometimes called the electra dolphin—and Fraser's dolphin *(Lagenodelphis hosei)* are also seldom with birds, but yet they are like the bird-associated spotted and spinner dolphins in being diurnally active, fast moving, and in large schools ($\bar{x}$ = 194 and 391, respectively), often mixed together. A sample from all research ship sightings between 1976 and 1981 that included all schools that were at least very likely either *Peponocephala* or *Feresa attenuata* (pygmy killer whale)—two species difficult to distinguish—gave twenty-three *Lagenodelphis* schools, of which only 13 percent were with bird flocks, and forty-two either *Peponocephala* or *Feresa* schools, of which only 7 percent were with birds (see bracketed results in table 5.3). It is clear, therefore, that birds are not strongly associated with any of these three dolphins.

Associations with Whales

We sighted many species of whales, and found they are generally not associated with birds (table 5.3). The rorquals we observed included blue *(Balaenoptera musculus)*, minke *(B. acutorostrata)*, sei *(B. borealis)*, and Bryde's *(B. edeni)* whales; these occurred singly or in small groups, all rarely with flocks. The ziphiid, or beaked whales, were also in small groups and were never seen with birds. Sperm whales *(Physeter macrocephalus)*, occasionally in large groups, were only rarely with birds. The related pygmy/dwarf sperm whales *(Kogia* spp.) were not seen with birds.

■

Seabirds, Dolphins, and Yellowfin Tuna

The Predominance of Spotted/Spinner Dolphins with Seabirds

Clearly, only spotted and spinner dolphins in the ETP are commonly associated with seabirds, an indication of underlying behavioral interactions that are very species specific. This predominance is made clear in figure 5.2, which shows the percentage of schools of the different dolphins that were with flocks, as well as the percentage of all associated flocks occurring with each dolphin species. Separate histograms are given for latitudes 5°N to 30°N (Tropical Water) and < 5°N to 12°S (Equatorial and Southern Subtropical waters). Among the spotted and spinner dolphins, in mixed or in pure schools, schools with spotted dolphins were usually most often with birds, particularly in the 5°N to 30°N Tropical Water.

The relationship of birds to spotted dolphins appears to be stronger than that of birds to spinner dolphins. Of the 242 flocks in table 5.3, 63.2 percent were associated with at least some spotted dolphin as opposed to 43.0 percent with at least some spinner dolphin. Also, 34.7 percent of these flocks were with unmixed spotted dolphin but only 10.3 percent with unmixed spinner dolphin schools. Most spinner dolphin schools associated with birds (78 percent of 113 = 88 schools; table 5.3) were also with spotted dolphins (76.1 percent of 88). Flocks were less likely to be with unmixed spinner dolphin schools in northern tropical waters and more likely in waters to the south (figure 5.2). This was because most bird-associated spinner schools in the northern waters were also with the spotted dolphin.

While this strong relationship of seabird flocks to spotted and spinner dolphins is characteristic of the Tropical Water habitat north of the equator, and also of the area south of the Galapagos Islands during the southern summer (December to February), these dolphins are relatively infrequently with birds elsewhere (compare the two latitude intervals in figure 5.2). This is so particularly in areas southwest of the Galapagos Islands, in spite of the abundance of flocks there. Table 5.4 shows the reduction in equatorial and southern latitudes of both the schools of these two dolphins and the percent of these schools that were associated with flocks of more than ten birds.

The Yellowfin Tuna Link

Tuna Under Flocks

Seabirds in flocks appear to feed mainly on prey driven to the surface by tunas, which are the only sizable pelagic fishes known to form abundant and large surface schools in the tropical ocean. Not infrequently we have been able to identify the predatory fish under

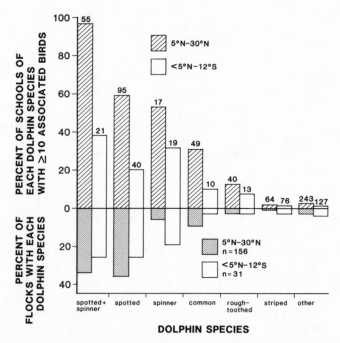

Figure 5.2. The rate of association of different dolphins with bird flocks (above) and the percent of these dolphin-associated flocks with each dolphin species (below). Numbers over the bars are the schools examined.

Table 5.4. Areawide differences in the percentage of spotted and spinner dolphin schools with bird flocks.

Center of Latitude Intervals	COAST TO 125°W			>125°W TO 155°W			TOTALS		
	Schools	Schools with flocks		Schools	Schools with flocks		Schools	Schools with flocks	
		n	%		n	%		n	%
20°N	9	3	33				9	3	33
15°N	76	51	67				76	51	67
10°N	50	33	66	11	5	46	61	38	62
5°N	67	32	48	3	3	100	70	35	50
0°	28	3	11	3	1	33	31	4	13
5°S	15	3	20	1	0	0	16	3	19
10°S	6	5[a]	83	3	0	0	9	5	56
Total	251	130		21	9		272	139[b]	

NOTES: Data from research cruises in 1976, 1977, 1979, 1980; some schools near land are deleted; flocks are of size $\geq$ 10 birds. Each latitude interval is defined by its center latitude $\pm$ 2.5°.

[a] These associated flocks were encountered south of the Galapagos Islands and west of Peru, where the porpoise-tuna fishery extends during the southern summer.

[b] 58.8% of schools in latitude intervals $\geq$ 5°N ($\pm$ 2.5°) were with flocks of size $\geq$ 10; only 21.4% of schools in latitude intervals < 5°N ($\pm$ 2.5°) were with such flocks. This differential is significant (X^2, $P < 0.005$).

a feeding flock as tuna (see also Murphy and Ikehara 1955; Ashmole and Ashmole 1967). The strongest evidence linking birds and dolphins to tuna is the existence of the "porpoise-tuna" fishery of the eastern Pacific, in which purse seiners catch yellowfin and to a lesser extent skipjack *(Katsuwonus pelamis)* tuna that swim with dolphins. These tuna, and the associated dolphins, are found mainly by searching the horizon for birds. Spotted, spinner, and common dolphins (ranked by importance) are the primary cetaceans involved in that fishery (Hammond 1981; Smith 1979). These are the same species, and order of importance, of dolphins that are associated with bird flocks, indicating that birds and dolphins are associated because of a tuna relationship.

The "Porpoise-Tuna" Fishery
The extent of the fishing grounds for surface-caught yellowfin is shown in figure 5.3 with the distribution of sightings of spotted dolphin superimposed (the distribution of spinner dolphin is nearly the same). The similarity of these distributions indicates an intimate species interaction, although such a result could be an artifact of joint fishing on both species by the purse seiners that supplied the data to the SWFC. However, the general pattern of the dolphin distribution has been confirmed by fishery-independent surveys (Au and Perryman 1985). Dolphin-associated yellowfin tend to be large ($\bar{x} = 120$ cm, s = 22 cm, years 1981–85) (see also Allen 1985) with sizes overlapping into the range of smaller yellowfin not caught with dolphins and of larger, deep-dwelling yellowfin that are caught by longline gear. Appendix 5.2 shows the relationship between tons of yellowfin caught and sizes of the associated dolphin schools on the purse seine grounds. The porpoise-tuna fishery indicates that seabirds and dolphins are linked via a relationship (probably feeding) with large yellowfin tuna.

■
The Ecology of Flocks

Species in Dolphin-Associated Flocks

Composition
The species composition of dolphin-associated flocks shows the same regionwide differences among flocks in general. The impor-

tant bird species that make up flocks associated with spotted and spinner dolphins are shown in figure 5.4, where again the histogram is separated for latitudes 5°N to 30°N and 5°N to 12°S. As in table 5.2, boobies, wedge-tailed shearwaters, jaegers, sooty terns, and frigate birds were numerically most important, in that order, in the dolphin-associated flocks from the northern Tropical Water habitat (5°N–30°N).

In the southern waters (< 5°N–12°S) boobies and sooty terns predominated; however, they were not usually in the same flocks. Boobies occurred mainly in flocks not too distant from coasts, while the sooty tern—more usually in virtually single-species flocks not with dolphins—was characteristic of the far offshore flocks (tables 5.1, 5.2).

Species Associations

Boobies, wedge-tailed shearwaters, jaegers, and frigate birds were found to be positively associated in dolphin-associated flocks from the Tropical Water purse seine fishing grounds, while frigate birds and sooty terns were positively associated in mainly the outer areas

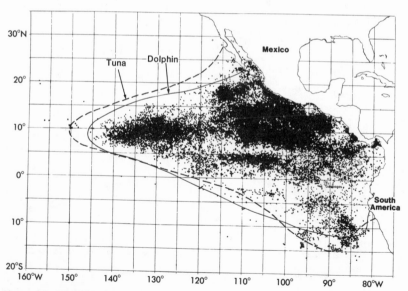

Figure 5.3. The distribution of records of spotted dolphins (dots and solid boundary, after Perrin et al. 1983) in relationship to the yellowfin tuna purse seiner grounds (dashed boundary, after Calkin's [1975] map of areas that produced 25+ tons of yellowfin catch).

of that habitat (Au and Pitman 1986). Table 5.5 is an extract of that analysis, showing percent co-occurrence of different bird species both in those flocks in which the species occurred, and in all flocks. The significant positive associations are also indicated. A schematic interpretation of this association complex is given in figure 5.5, where two measures of strength of association between species pairs are presented: Cole's (1949) coefficient and Yule's contingency index (see Pielou 1969:164). Our data indicate that none of the associations are very strong. The figure suggests that frigate birds are positively associated with other species in all habitats. The simplest explanation of these positive associations is attraction to common feeding opportunities.

Feeding Tactics/Strategies

Multispecies Flocks and Facultative Commensals
    Within multispecies flocks, simultaneously different, species-specific feeding behaviors suggest prey aggregations that are diverse

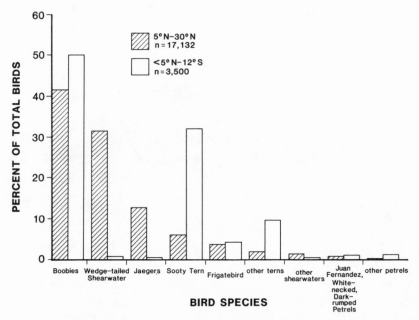

Figure 5.4. Likelihood (percentage frequency) of finding bird species in flocks associated with spotted and spinner dolphins.

Table 5.5. Percentage of co-occurrence of species pairs in flocks.

		Boobies n = 111	Frigate Birds n = 90	WTSW n = 82	Large Pterodromas n = 18	Jaegers n = 80
Sooty tern	% n	45	53*	49	56	46
	(% total)	(38)	(37)*	(31)	(8)	(28)
Boobies	% n		90*	90*	44	92*
	(% total)		(62)*	(56)*	(6)	(56)*
Frigate birds	% n			72	50	75
	(% total)			(45)	(7)	(46)
WTSW	% n				50	74*
	(% total)				(7)	(45)*
Large Pterodromas	% n					11
	(% total)					(7)

NOTES: From n = 131 flocks associated with spotted or mixed spotted and spinner dolphin schools in latitudes 5°N–30°N in the eastern Pacific; example: sooty terns co-occurred with boobies in 45 percent of the 111 flocks containing boobies, and in 38 percent of all 131 flocks; * indicates a significant positive association between the species pair (by X^2, $P < 0.05$); there were no significant associations detectable among the thirty-three spotted/spinner dolphin–associated flocks from latitude < 5°N. WTSW = wedge-tailed shearwater.

with respect to behavior, distribution in the water column, size, or species composition (we will refer to this diversity as prey "configuration"). Boobies, flying rapidly back and forth, especially over the advancing front of a tuna school, wheel and plunge after their prey or make midair captures. Wedge-tailed shearwaters seemingly race with the boobies, but at lesser heights, and then drop to the water in surface plunges or for contact dipping or surface seizing of prey that appear available for at least several seconds. Jaegers, taking prey by aerial pursuit and dipping, and occasionally by piracy, add to the scene of frantic activity (feeding methods are defined by Ashmole [1971]). By rapidly covering the school, birds of each species increase their encounter rate with unpredictably available and fleeting prey. Sooty terns employ another tactic—they watch widely and deep into the water for developing feeding opportunities from positions high above. Their prey appear to be grouped, for the entire

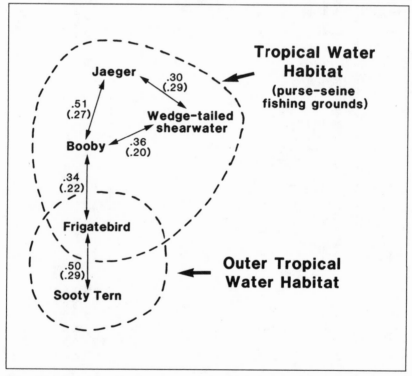

Figure 5.5. Positive species-association links within flocks, with strength-of-association coefficients (Cole's and Yule's [in parentheses] coefficients) shown.

flock will swoop down synchronously. Frigate birds soar high above the feeding melee and swoop down at opportune moments after individual prey that are likely larger than that of the terns.

Except for sooty terns and frigate birds, most birds that occur in these multispecies flocks are facultative commensals with tuna. When feeding independently, these birds occur singly or in small groups, and aerial feeding is rarely seen. Petrels and shearwaters, for example, then rely more on scavenging or preying upon free-floating organisms such as *Velella* (Pitman personal observations). Boobies, however, appear to feed independently of tuna the least, and their distribution most closely coincides with that of the yellowfin tuna fishery (cf. figures 5.3 and 5.6, considering also the low-density, westward extension of booby habitat [below]). Boobies are both the most abundant of birds in flocks over tuna and dolphins (figure 5.4) and the most abundant of seabirds in the northern Tropical Water (Pitman 1986), with colonies in the eastern Pacific that are probably the world's largest (Nelson 1978). The abundance of boobies is likely a direct consquence of their strong association with yellowfin.

"Obligate" Commensals

In contrast to the facultative commensals, sooty terns and frigate birds are almost never seen feeding independently of tuna in oceanic

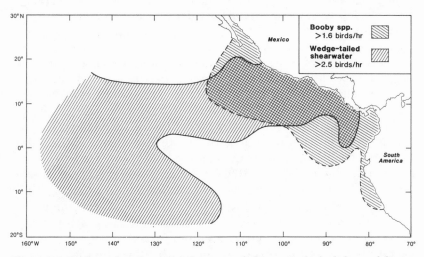

Figure 5.6. Booby and wedge-tailed shearwater habitats. Each shaded area delimits regions of greater than median sighting rate of the species (birds/hour). After Pitman 1986.

areas; they appear to be near-obligate commensals with these fish. The sooty tern, in particular, is most abundant beyond the yellow-fin fishing grounds, where the tuna it feeds with are generally small, probably skipjack (Hida 1970), but possibly also frigate mackerel (*Auxis* spp.; see Olson and Boggs 1986) or small yellowfin and bigeye tuna *(Thunnus obesus)*. It is the most abundant seabird of the southeastern and central Pacific (Pitman 1986; Gould 1974). In the former area, there is apparently little opportunity for feeding independently of tunas; facultative commensals (i.e., most other birds) are virtually excluded from those waters, apparently being unable to feed either with sooty terns or independently (table 5.2 and below). The obvious ecological success of the sooty tern probably stems from its ability to follow and feed with the small tuna.

Feeding Regimes

Species Distribution
Distinctly different distributions of four birds that occur in flocks illustrate how the community dichotomy is formed from seabirds that seem to feed differently. The distributions are depicted in figures 5.6 and 5.7, where a single contour is used to enclose areas of higher population density of each species. These were derived from distribution and relative abundance studies by Pitman (1986). Each species' contour approximately delimits areas where densities were greater than that species' median sighting rate (birds per hour). Figure 5.6 shows boobies and wedge-tailed shearwaters inhabiting the Tropical Water north of the equator, but seldom occurring in a large area of the southeastern tropical Pacific west of the Peru Current. The habitat of the wedge-tailed shearwater extends into the eastern Pacific from broad areas to the west, while that of boobies extends from the east, westward with the yellowfin tuna fishing grounds. The lower-density (less than median sighting rate) habitat of boobies (primarily the masked booby) extends far west of Clipperton Island—at 10°N, 109°W (Pitman 1986) (this is not shown clearly in figure 5.6 because areas of greater than median sighting rate of boobies, especially the red-footed booby, are compressed toward the American coasts). Figure 5.7 shows an extensive high-density area of the sooty tern habitat in the central Pacific, extending into the Subtropical Water of the southeastern Pacific, the area with a dearth of boobies and wedge-tailed shearwaters. The high-density areas of this tern are sparse in the Tropical Water north of

the equator. Finally, Juan Fernandez/white-necked petrels occur in a band about latitude 10°N, especially to the west of longitude 110°W.

Not shown is the jaeger habitat, mainly within a thousand-kilometer-wide band along the coast of Middle America, and the fact that the larger seabirds are relatively uncommon along the equator, being replaced there by plankton-feeding storm petrels. The distributions of the seabirds also vary seasonally; in particular, sooty terns probably extend farthest into the southeastern Pacific during the southern summer.

Regional Prey Differences

The northern (Tropical Water) and the southern (Equatorial and Southern Subtropical waters) habitats of the ETP thus appear to have different prey characteristics or configurations that require different foraging tactics, as indicated by community differences in seabird species and their interactions. In the northern Tropical Water, prey patches appear to be relatively large and to have diverse kinds of prey. Once found, hundreds of birds and dolphins (see table 5.3) and yellowfin (e.g., 10 tons of fish 60 lbs or greater; see IATTC 1984 and appendix 5.2) may feed upon the patch. Feeding many continue for some time (we have watched this activity for nearly an hour before continuing on). Under such conditions, satiated birds, e.g.,

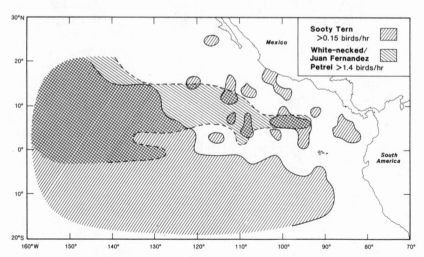

Figure 5.7. Sooty tern and white-necked/Juan Fernandez petrel habitats. Each shaded area delimits regions of greater than median sighting rate of the species (birds/hour). After Pitman 1986.

boobies, petrels, and especially shearwaters, may rest upon the water while others continue to feed (personal observation). In contrast, prey in the less productive southeastern and central Pacific appear to be in small, ephemeral, and thinly scattered patches of simple configuration, judging by the behavior and composition of feeding flocks. The buoyantly flying sooty tern especially, which is unable to rest long, or at all, upon the water, is evidently adapted for exploiting such prey, whose availability appears to change quickly, lending few opportunities for heavy feeding. These habitat differences probably reflect dissimilarities in biological productivity. The northern Tropical Water is richest because of its proximity to land and mechanisms that carry nutrients across its strong, shallow thermocline (Brandhorst 1958; Wyrtki 1966).

Pertubations

If prey characteristics select for particular bird-tuna-dolphin interactions, environmental perturbations could alter these behaviors through effects on forage. During the 1982–83 El Niño warming event in the Pacific (Philander 1983), fishermen experienced a 25 percent reduction in the yellowfin tuna catch (IATTC 1984) and reported fewer porpoise-tuna schools. Because El Niño episodes usually result in reduced biological production and a deepened thermocline (see Barber and Chavez 1983), a weakened tuna-dolphin association might be expected, as is characteristic of the deep-thermocline and less productive southeastern and central Pacific. We looked for such effects in the characteristics of flocks recorded off Middle America (from Baja California to the equator) in 1979, 1980, and 1983. Records of all flocks and of flocks associated with spotted, spinner, and common dolphins were examined. These data, summarized in table 5.6, suggest that flock density was similar over the years, as was species composition: boobies and wedge-tailed shearwaters were always most abundant. However, flock size, dolphin school size, the percent of dolphin schools with birds, and the ratio of size of dolphin-associated flocks to that of all flocks were all much reduced in 1983. Although the 1983 sample was probably too small to be representative, these reduced percentages in 1983 are consistent with the idea that decreased food production or availability near the surface results in tuna feeding more independently of dolphins and in ways that are less useful to birds.

■

General Discussion

It is curious that the association of seabirds, yellowfin tuna, and dolphins is so specific to spotted and spniner dolphins; the birds essentially ignore the twenty or so other species of Cetacea in the eastern Pacific. This is somehow related to the conspicuous partitioning of the eastern tropical Pacific (ETP) into distinct epipelagic communities of differently interacting predators.

Foraging Tactics and Species Interactions

Tuna Strategy and Consequences
To begin understanding oceanic species interactions, it is useful to consider the behavior of the surface-schooling tunas. Tuna behavior exemplifies a strategy for exploiting the relatively sparse prey of tropical seas. Tropical tunas feed on epipelagic fish, squids, and crustaceans, whose distributions are undoubtedly very patchy with low overall densities (Blackburn 1968). To ensure a sufficient capture rate of such prey, tunas search by extensive ranging in the horizontal dimension, an energetically demanding tactic. Norberg (1977) postulated that as prey density decreases, the search method required of a predator increasingly becomes both more energy consuming and more efficient. Thus tunas have evolved into perhaps the most streamlined of fishes; their whole morphology and physiology appear designed for fast, sustained swimming with metabolic rates probably higher than those of all other fishes (see Sharp and Dizon 1978 for descriptions of energetics and hydrodynamics). High energy expenditure to obtain moderate energy returns from low-density prey must constrict the tunas' positive energy balance, narrowing the scope, or margin, between energy gained from food and energy consumed (see Warren and Davis 1966). Constriction of this energy margin increases vulnerability to natural mortality; for their size and speed, tropical tunas are remarkably short-lived, living probably less then ten years on the average (Beverton and Holt 1959).

These costs notwithstanding, the efficacy of the tunas' feeding strategy is evident; they are the dominant pelagic fishes of the tropical ocean, supporting extensive bird populations and productive fisheries. During the period 1974 to 1981 (between major El Niño events) an average of 196,000 metric tons of yellowfin tuna

200   David W. Au and Robert L. Pitman

was harvested annually from the ETP, virtually all from surface or
near-surface schools (IATTC 1984), approximately half of which
were dolphin-associated (Allen 1985). In the same period, skipjack
tuna produced an average annual catch of 122,000 metric tons,
mainly from "schoolfish," i.e., surface schools without dolphins.

Feeding with Tuna
   Seabirds that feed with surface tunas exploit the tunas' tactic for
finding prey and their habit of driving it to the surface. This role
was recognized by Ashmole and Ashmole (1967) and is the basis for
fishermen's reliance on birds to locate these fish. The degree to
which tuna schools are accompanied by flocks, however, is difficult
to estimate, as most schools are detected only if associated with
birds. Appendix 5.3 shows that whereas most schools fished by
seiners were with birds (as expected), schoolfish tuna were both less
often and more variably with birds (29 percent to 77 percent overall
depending upon area).
   Dolphins that associate with yellowfin tuna also may have a
feeding relationship with these fish. The prey of the spotted dolphin
is similar to that of yellowfin, both feeding diurnally upon epipelagic fishes and squids (Perrin et al. 1973; Reintjes and King 1953).
Seabirds feed on much the same kinds of prey (Harrison, Hida, and
Seki 1983; Diamond 1983). Spinner dolphins feed more on mesopelagic animals and may be less directly linked to tuna (Perrin et al.
1973). The ecological success of spinner and especially spotted dolphins in the eastern Pacific, which may be appreciated by comparing the numbers and average sizes of their schools with that of the

Table 5.6. Comparison of flock characteristics and interactions off middle
America during three years.

Year	TOTAL Flocks	TOTAL Birds	Hours observed	Flocks per hour	% Flocks with dolphins
1979	136	10,366	270.0	0.5	27
1980	73	3,721	191.6	0.4	45
1983	25	800	49.0	0.5	28

NOTE: Dolphins referred to are spotted, spinner, and common dolphins.
WTSW = wedge-tailed shearwater.

other cetaceans (table 5.3), is likely due to a considerable advantage gained from feeding with yellowfin tuna.

Seabirds' Relationship to Cetaceans

In spite of the strong association of spotted and spinner dolphins with birds, our observations indicate that cetaceans themselves have little role in making prey available to seabirds. It is the tuna, primarily, that appear to drive prey to the surface, whether or not dolphins are present; foraging birds are almost always most active where the tuna are feeding, which often is at the leading edge of the school. This has been seen from ships (personal observation) and from helicopters (Au and Perryman 1982; Hewitt and Friedrichsen ms.). Thus spotted and spinner dolphins are commonly with birds because they swim with tuna with which the birds feed. And the common dolphin is relatively infrequently found with flocks because its association with tuna is similarly infrequent. Though the rough-toothed dolphin is not uncommonly associated with small flocks, it was probably the flotsam (and associated fish) near which this species was often encountered, rather than tuna or the mammals, that attracted the birds. Tropical seabirds thus appear to strongly associate only with those dolphins that swim and feed with tuna, dolphins that, like the tuna, are fast traveling and in large, diurnally active schools. Even so, these tuna and birds appear to feed independently of dolphins much of the time, as indicated by the delphinid association rate of flocks: between latitudes 2.5°N and 22.5°N, 58 percent of flocks (presumably with tuna) were not with dolphins (data for table 5.1, Eastern Sector). The tuna-dolphin association is

PERCENT SPP. COMPOSITION OF FLOCKS								FLOCK SIZE		DOLPHIN SCHOOLS
Boobies	Sooty tern	Other tern	WTSW	Other shear- waters	Jaegers	Frigate birds	Other birds	All	With dolphins	Size	Pct. With birds
38	13	3	34	1	8	2	1	76.2	159.8	297.9	75
23	17	12	15	1	11	7	3	51.0	88.4	149.2	76
16	1	10	66	2	4	2	0	32.0	21.7	115.2	29

clearly not obligatory; it seems rather to indicate an intersection or overlap of certain foraging tactics adopted by these species to exploit local prey configurations.

Communities and Foraging Requirements

The foraging tactics required in a particular environment seem to shape communities by placing stringent demands upon the behavioral or energetic capabilities of predators, especially in biologically sparse waters. Thus while spotted and spinner dolphins regularly associate with large yellowfin tuna on the purse seine fishing grounds of the Tropical Water habitat, these tuna are largely unable to forage in the surface layer beyond those grounds, perhaps in part due to physiological needs (see Sharp 1978 for a detailed discussion), and the dolphins there seem unable to feed with the skipjack or other small tuna that replace the yellowfin. The multispecies flocks of both facultative and obligatory commensals that feed with yellowfin on the purse seine grounds are reduced to mainly sooty terns and wedge-tailed shearwaters in the biologically sparse waters of the central Pacific and to virtually sooty tern–only flocks, obigatorily commensal on small tuna, in the Southern Subtropical Water. Like other facultative commensals, wedge-tailed shearwaters, so widespread in the central Pacific (Pitman 1986; King 1974b), are apparently unable to assume the required feeding tactics and to penetrate the Southern Subtropical Water of the eastern Pacific (figure 5.6).

Who Follows Whom?

It is clear that seabirds follow and benefit from feeding with tuna, but where these tuna are also associated with certain dolphins, does either the tuna or dolphin provide benefit to the other? This question is pertinent to understanding the role of dolphins in bird-tuna-dolphin associations. A widely held view is that tuna follow dolphins in foraging. Tuna evidently do follow dolphins that are chased by purse seiners and are eventually captured with these schools. Mullen (1984) showed how two potentially competing predators could theoretically and stably coexist in a commensal relationship and suggested that the tuna were commensals on the dolphins. But perhaps tuna, and pelagic schooling fishes in general, obtain protection from pursuing predators by crowding under objects they normally encounter, both animate and inanimate—as when a predacious marlin approaches, skipjack will hide under the boat fishing them (D. Correa, personal communication). We suggested (Au and Pitman 1986) that spotted and spinner dolphins follow yellowfin in jointly foraging schools, inferring this mainly because seabirds are often seen feeding at the front of such schools in immediate association with the tuna rather than the dolphins. We note that the tuna-dolphin association breaks down outside the ETP purse seine grounds, where large yellowfin no longer commonly school at the surface (the dolphins must continue to do so, though they are no longer often with birds), and that a large tuna school, searching in three-dimensional space and perhaps using olfactory cues, could be more efficient than dolphins in locating prey.

In retrospect, it seems unlikely that a simple answer, either the dolphins or the tuna following the other to food, could be satisfactory. Neither is it likely that the tuna-dolphin-bird association merely results from the convergence of predator species upon the same food patches, for the association is too species specific, and tunas and dolphins appear to travel together, even while not actively feeding. Finally, no explanation can be satisfactory unless it also explains why the dolphin-tuna association is not characteristic of the eastern tropical Atlantic (Levenetz, Fonteneau, and Regalado 1980; Stretta and Slepoukha 1986), where a large purse seine fishery for surface yellowfin and skipjack tuna also exists, and dolphins similar to that of the ETP occur (Leatherwood, Caldwell, and Winn 1976).

A Feeding Tactics Hypothesis

Hypothesis

We propose a hypothesis that particular prey configurations or arrangements that are a function of productivity shape species interactions through the foraging tactics required to exploit that prey. The resulting explanation of species interactions is as follows: In low productivity waters, low-density prey are exploited by skipjack or similar small-sized tuna that are specialists at surviving on small prey from highly dispersed, relatively small patches. These patches are sufficiently encountered only through the most rapid and energetically expensive, wide-ranging search. Under such conditions smaller, rather than larger, predators are at an advantage (Norberg 1977). In accordance with foraging theory (Charnov 1976), these tuna employ, in effect, hit-and-run tactics on patches not much more profitable than from average searching between patches. Ecologically successful seabirds in such areas would be those capable of keeping up with these fast tuna, there being little food available independent of the fish.

In areas of intermediate food productivity, such as along oceanic boundary zones, food patches and prey are larger, though still best discovered by extensive horizontal ranging. Once found, these patches often provide for relatively long feeding bouts. Large yellowfin tuna and similarly foraging spotted and spinner dolphins find and exploit this prey, often jointly. When doing so, the tuna may be the primary predator, and they may drive some prey to the surface. Many birds take advantage of these enhanced feeding opportunities, forming multispecies flocks; however, most species in these flocks can also supplement their food by also feeding independently of tuna.

In still higher-productivity waters, such as coastal areas, prey is more diverse, the food encounter rate is high, prey patches are large and more predictable, and the advantage of the wide-ranging foraging tactic of tunas is lessened. It may sometimes be advantageous for tunas to forage passively—for example by waiting for prey under objects. The richer and less clumped food resources would enable the different predators to specialize and to feed more independently.

Extensions

The hypothesis would predict that reductions in food productivity would reduce the participation of all species involved in joint feeding, as suggested for 1983 in table 5.6 (this would not be expected if there were an obligate commensal relationship between tuna and dolphins). Moreover, the switching of feeding tactics and hence changes in community interactions might be the mode of response to such changes. Such a mechanism may have been involved in the massive population failure of seabirds from Christmas Island during the last El Niño (see Schreiber and Schreiber 1984).

The hypothesis suggests that intermediate productivity would be most conducive to the formation of multispecies interactions involving birds, tuna, and dolphins. In fact, the most extensive porpoise-tuna fishing areas in the eastern Pacific are not the rich coastal and upwelling-influenced waters off Central America, but the warm, stable waters off southern Mexico and the waters west of Clipperton Island. Areas where two-thirds or more of purse seine operations are on dolphin-associated yellowfin begin about 600 kilometers offshore (see Allen 1985), except off southern Mexico, where the fishing comes close to shore. Could it be that yellowfin tuna in the eastern tropical Atlantic are seldom associated with dolphins because waters there are not sufficiently intermediate in productivity to develop the right prey configuration? That fishery is located mainly in the Equatorial Counter and Guinea currents within 500 kilometers of the southern West African coast, an environment more similar to Pacific waters west of Panama and Colombia than to the ridged thermocline (Cromwell 1958), offshore porpoise-tuna grounds west and southwest of southern Mexico (see Merle 1978).

Background

This idea, that the resource base in different environments controls species interactions through the tactics required for its exploitation, is patterned after the concepts developed by Crook (1965) in a study on birds and as applied in comparative behavior studies of primates (Crook 1970; Clutton-Brock and Harvey 1977). These authors explained how the social organization and behavior of species are shaped by the availability of food and sites for reproduction. Smith et al. (1986), noting that common dolphins tended to occur in large, mobile schools in oceanic waters off California, while Dall's porpoise *(Phocoenoides dalli)* occurred in small, relatively sedentary schools in chlorophyll-rich, coastal waters, suggested that the behavioral and population differences were due to feeding strategies required of each species in the different environments. Differences in feeding behaviors of seabirds from Antarctic to tropical seas were explained by Ainley and Boekelheide (1983) as adaptations to regional differences in prey density and patchiness, as well as to the presence of subsurface competitors and predators. Wiens (1984) reviewed the importance of resources in the organization of avian populations and communities, noting how little direct evidence there was of resource limitation. He cautioned against imposing possibly preconceived processes, such as competition, on the analysis of relationships. However, Safina and Burger (1985) thought that terns in coastal waters could compete with bluefish by pursuing the same individual prey. We have not directly considered competition as a mechanism in our hypothesis. Until demonstrated, Schoener's hypothesis (1982) of predator convergence onto locally abundant prey patches, with little interspecies competition, seems more likely. Ours is an attempt to explain the existence in the eastern tropical Pacific of distinct pelagic communities, not separated by physical barriers, and composed of specific assemblages of highly mobile species with specific behavioral interactions.

■
Summary

Two distinct faunal communities involving seabirds may be recognized in the eastern tropical Pacific. One, characteristic of the Tropical Water habitat mainly north of latitude 5°N, consists of multispecies flocks of primarily boobies, wedge-tailed shearwaters, jaegers, and sooty terns. These flocks are frequently associated with large yellowfin tuna and dolphins. Of the dolphins, the spotted and spinner species predominate (ca. 75 percent of dolphin-associated flocks involve these two species). These dolphins appeared to be linked to birds because both feed with large yellowfin tuna wherever the latter forage close to the surface. The other community occurs primarily in Subtropical Water to the south and consists of virtually single-species flocks of sooty terns. These flocks are associated with small tuna but seldom with dolphins. We propose that the different kinds of species associations seen in the eastern Pacific are manifestations of different foraging tactics required of pelagic predators in the different areas, and that the intersection of such tactics could explain the bird-tuna-dolphin association.

Acknowledgments

For critical and constructive reviews, we thank Norman W. Bartoo, Gary T. Sakagawa, Andrew E. Dizon, William F. Perrin, and Robert J. Olson. We profited from critiques and discussions with Joseph Jehl, Ralph W. Schreiber, and David G. Ainley in connection with an earlier version of this paper. We thank Henry Orr for drafting the illustrations and Joan Michalski and Lorraine C. Prescott for final editing and typing.

References

Ainley, D. G. and R. J. Boekelheide. 1983. An ecological comparison of oceanic seabird communities of the South Pacific Ocean. In R. W. Schreiber, ed., *Tropical Seabird Biology*, pp. 2–23. Cooper Ornithological Society, Studies in Avian Biology, no. 8.

Allen, R. L. 1985. Dolphins and the purse-seine fishery for yellowfin tuna. In J. R. Beddington, R. J. H. Beverton, and D. M. Lavigne, eds., *Marine Mammals and Fisheries*, pp. 238–252. Winchester, Md.: Allen and Unwin.

Ashmole, N. P. 1971. Seabird ecology and the marine environment. In D. S. Farner, D. R. King, and K. C. Parkes, eds., *Avian Biology*, 1:223–286. New York and London: Academic Press.

Ashmole, N. P. and M. J. Ashmole. 1967. Comparative feeding ecology of seabirds of a tropical oceanic island. *Peabody Mus. Nat. Hist., Yale Univ. Bull.* 24:1–131.

Au, D. W. K. and W. L. Perryman. 1982. Movement and speed of dolphin schools responding to an approaching ship. *U.S. Fish. Bull.* 80(2):371–379.

Au, D. W. K. and W. L. Perryman. 1985. Dolphin habitats in the eastern tropical Pacific. *U.S. Fish. Bull.* 83(4):623–643.

Au, D. W. K. and R. L. Pitman. 1986. Seabird interactions with dolphins and tuna in eastern tropical Pacific. *Condor* 88(3):304–317.

Barber, R. T. and F. P. Chavez. 1983. Biological consequences of El Niño. *Science* 222(4629):1203–1210.

Beverton, R. J. H. and S. J. Holt. 1959. A review of lifespan and mortality rates of fish in nature, and their relation to growth and other physiological characteristics. In G. E. W. Wolstenholme and M. O'Connor, eds., *The Lifespan of Animals, Ciba Foundation Colloquia on Aging*, 5:142–177. London: J. and A. Churchill.

Blackburn, M. 1968. Micronekton of the eastern tropical Pacific Ocean: family composition, distribution, abundance, and relations to tuna. *U.S. Fish. Bull.* 67:71–115.

Brandhorst, W. 1958. Thermocline topography, zooplankton standing crop, and mechanisms of fertilization in the eastern tropical Pacific. *J. Cons. Int. Explor. Mer* 24:16–31.

Calkins, T. P. 1975. Geographic distribution of yellowfin and skipjack tuna catches in the eastern Pacific Ocean, and fleet and total catch statistics, 1971–1974. *Inter-American Tropical Tuna Comm., Bull.* 17:3–116.

Cole, L. C. 1949. The measurement of interspecies association. *Ecology* 30:411–424.

Charnov, E. L. 1976. Optimal feeding: the marginal value theorem. *Theor. Popul. Biol.* 9:129–136.

Clutton-Brock, T. H. and P. H. Harvey. 1977. Primate ecology and social organization. *J. Zool., London,* 183:1–39.

Cromwell, T. 1958. Thermocline topography, horizontal currents and "ridging" in the eastern tropical Pacific. *Inter.-Amer. Tropical Tuna Comm., Bull.* 3:135–164.

Crook, J. H. 1965. The adaptive significance of avian social organization. *Symp. Zool. Soc. London,* 14:181–218.

Crook, J. H. 1970. The socio-ecology of primates. In J. H. Crook, ed., *Social Behavior in Birds and Mammals*, pp. 103–166. London: Academic Press.

Diamond, A. W. 1983. Feeding overlap in some tropical and temperate seabird communities. In R. W. Schreiber, ed., *Tropical Seabird Biology*, pp. 24–46. Cooper Ornithological Society Studies in Avian Biology no. 8.

Fitch, J. E. and R. L. Brownell. 1968. Fish otoliths in cetacean stomachs and their importance in interpreting feeding habits. *J. Fish. Res. Bd. Canada* 25:2561–2674.

Gould, P. J. 1971. *Interactions of seabirds over the open ocean.* Ph.D. thesis. University of Arizona (Tucson).

Gould, P. J. 1974. Sooty Tern *(Sterna fuscata).* In W. B. King, ed., *Pelagic Studies of Seabirds in the Central and Eastern Pacific Ocean*, pp. 6–52. Smithsonian Contributions to Zoology, no. 158.

Hammond, P. S., ed. 1981. *Report on the workshop on tuna dolphin interactions, Managua, Nicaragua. Inter-American Trop. Tuna Comm.* Special Report no. 4.

Harrison, C. S., T. S. Hida, and M. P. Seki. 1983. Hawaiian seabird feeding ecology. *Wildl. Monogr.* no. 85.

Hewitt, R. P. and G. L. Friedrichsen. Manuscript. On birds and dolphins. Southwest Fisheries Center, La Jolla, Calif.

Hida, T. S. 1970. Surface tuna schools located and fished in equatorial eastern Pacific. *Comm. Fish. Rev.* 32(4):34–37.

Hill, M. O. 1973. Diversity and evenness: a unifying notation and its consequences. *Ecology* 54(2):427–432.

IATTC. 1984. *Annual Report of the Inter-American Tropical Tuna Commission* (IATTC), La Jolla, Calif. Table 1.

King, W. B. 1970. The trade wind zone oceanography pilot study, part 7: Observations of Seabirds, March 1964 to June 1965. Washington, D.C.: U.S. Fish and Wildlife Service, Special Scientific Report, Fisheries 586.

King, W. B., ed. 1974a. *Pelagic Studies of Seabirds in the Central and Eastern Pacific Ocean*, pp. 53–95. Smithsonian Contributions to Zoology, no. 158.

King, W. B. 1974b. Wedge-tailed Shearwater *(Puffinus pacificus)*. In W. B. King, ed., *Pelagic Studies of Seabirds in the Central and Eastern Pacific Ocean*, pp. 53–95. Smithsonian Contrib. to Zool. no. 158.

Leatherwood, S., D. K. Caldwell, and H. E. Winn. 1976. *Whales, Dolphins, and Porpoises of the Western North Atlantic.* NOAA Technical Report NMFS CIRC-396, National Marine Fisheries Service, Seattle, Wash.

Levenetz, J., A. Fonteneau, and R. Regalado. 1980. Resultats d'une enquete sur l'importance des dauphins dans la pecherie thoniere FISM. International Commission for the Conservation of Atlantic Tunas, Collective Volume of Scientific Papers 9(1):176–179.

Merle, J. 1978. *Atlas hydrologique saisonnier de l'ocean Atlantique intertropical.* Travaux et documents de l'Office de la Recherche Scientifique et Technique Outre-Mer, no. 82.

Miyazaki, N., T. Kasuya, and M. Nishiwaki. 1973. Food of *Stenella coeruleoalba. Sci. Rep. Whales Res. Inst.* 25:265–275.

Mullen, A. J. 1984. Autonomic tuning of a two predator, one prey system via commensalism. *Math. Biosciences* 72:71–81.

Murphy, G. I. and I. I. Ikehara. 1955. *A Summary of Sightings of Fish Schools and Bird Flocks and of Trolling in the Central Pacific.* Washington, D.C.: U.S. Fish and Wildlife Special Scientific Report, Fisheries 154.

Nelson, J. B. 1978. *The Sulidae: Gannets and Boobies.* Oxford: Oxford University Press.

Norberg, R. A. 1977. An ecological theory on foraging time and energetics and choice of optimal food-searching method. *J. Anim. Ecol.* 46:511–529.

Olson, R. J. and C. H. Boggs. 1986. Apex predation by yellowfin tuna *(Thunnus albacares):* Independent estimates from gastric evacuation and stomach contents, bioenergetics, and cesium concentrations. *Canad. J. Fish. Aquat. Sci.* 43(9):1760–1775.

Perrin, W. F., M. D. Scott, G. J. Walker, F. M. Ralston, and D. W. K. Au. 1983. *Distribution of Four Dolphins* (*Stenella* spp. and Delphinus delphis) *in the Eastern Tropical Pacific, with an Annotated Catalog of Data Sources.* National Marine Fisheries Service, La Jolla, California: National Oceanic and Atmospheric Administration Technical Memorandum NMFS-SWFC-38.

Perrin, W. F., R. R. Warner, C. H. Fiscus, and D. B. Holts. 1973. Stomach contents of porpoise, *Stenella* spp., and yellowfin tuna, *Thunnus albacares*, in mixed species aggregations. *U.S. Fish. Bull.* 71(4):1077–1092.

Philander, S. G. H. 1983. El Niño southern oscillation phenomena. *Nature* 302:295–301.

Pielou, E. C. 1969. *An Introduction to Mathematical Ecology.* New York and London: Wiley Interscience.

Pitman, R. L. 1986. *Atlas of Seabird Distribution and Relative Abundance in the*

Eastern Tropical Pacific. Southwest Fisheries Center, National Marine Fisheries Service, National Oceanic and Atmospheric Administration, Administrative Report LJ-86-02C.

Reintjes, J. W. and J. E. King. 1953. Food of the yellowfin tuna in the central Pacific. *U.S. Fish. Bull.* 54:91–110.

Robinson, B. H. and J. E. Craddock. 1983. Mesopelagic fishes eaten by Fraser's dolphin, *Lagenodelphis hosei. U.S. Fish. Bull.* 81(2):283–289.

Safina, C. and J. Burger. 1985. Common tern foraging: seasonal trends in prey fish densities and competition with bluefish. *Ecology* 66(5):1457–1463.

Schoener, T. W. 1982. The controversy over interspecific competition. *Amer. Sci.* 70(6):516–595.

Schreiber, R. W. and E. A. Schreiber. 1984. Central Pacific seabirds and the El Niño Southern Oscillation: 1982–1983 perspective. *Science* 225(4663):713–716.

Sharp, G. D. 1978. Behavioral and physiological properties of tunas and their effects on vulnerability to fishing gear. In G. D. Sharp and A. E. Dizon, eds., *The Physiological Ecology of Tunas,* pp. 397–449. New York: Academic Press.

Sharp, G. D. and A. E. Dizon. 1978. *The Physiological Ecology of Tunas.* New York: Academic Press.

Smith, R. C., P. Dustan, D. Au, K. S. Baker, and E. A. Dunlop. 1986. Distribution of cetaceans and sea surface chlorophyll concentration in the California Current. *Marine Biology* 91(3):385–402.

Smith, T. D., ed. 1979. *Report of the Workshop on Status of Porpoise Stocks, La Jolla, Calif., August 27–31, 1979.* Southwest Fisheries Center, National Marine Fisheries Service, Administrative Report LJ-70-41.

Stretta, J. M. and M. Slepoukha. 1986. Analyse des facteurs biotiques et abiotiques associes aux bancs de thons. In P. E. K. Symons, P. M. Miyake, and G. T. Sakagawa, eds., *Proceedings, ICCAT Conference on the International Skipjack Year Program,* International Commission for the Conservation of Atlantic Tunas, pp. 161–169. Madrid: ICCAT.

Warren, C. E. and G. E. Davis. 1966. Laboratory studies on the feeding, bioenergetics, and growth of fish. In S. D. Gerking, ed., *The Biological Basis of Freshwater Fish Production,* pp. 175–214. Oxford and Edinburgh: Blackwell Scientific.

Wiens, J. A. 1984. Resource systems, populations, and communities. In P. W. Price, C. N. Slobodchikoff, and W. S. Gaud, eds., *A New Ecology—Novel Approaches to Interactive Systems,* pp. 397–436. New York: Wiley.

Wyrtki, K. 1964. Upwelling in the Costa Rica Dome. *U.S. Fish. Bull.* 63:355–372.

Wyrtki, K. 1966. Oceanography of the eastern equatorial Pacific Ocean. *Oceanogr. Mar. Biol. Ann. Rev.* 4:36–68.

Appendix 5.1. Common and scientific names of species mentioned in text.

TUNAS

Yellowfin	*Thunnus albacares*
Bigeye	*T. obesus*
Skipjack	*Katsuwonus pelamis*
Frigate mackerel	*Auxis* spp.

BIRDS

Boobies	
Red-footed	*Sula sula*
Masked	*S. dactylatra*
Brown	*S. leucogaster*
Wedge-tailed shearwater	*Puffinus pacificus*
Sooty tern	*Sterna fuscata*
Jaegers	*Stercorarius* spp.
Juan Fernandez petrel	*Pterodroma externa externa*
White-necked petrel	*Pterodroma externa cervicalis*
Dark-rumped petrel	*Pterodroma phaeopygia*
White tern	*Gygis alba*
Noddy terns	*Anous* spp.
Frigatebirds	*Fregata* spp.
Phalaropes	(Phalaropodidae)
Storm-petrels	(Hydrobatidae)
Gulls	*Larus* spp.
Tropic birds	*Phaethon* spp.

CETACEANS

Delphinids	
Spotted dolphin	*Stenella attenuata*
Spinner dolphin	*S. longirostris*
Striped dolphin	*S. coeruleoalba*
Common dolphin	*Delphinus delphis*
Bottlenose dolphin	*Tursiops truncatus*
Risso's dolphin	*Grampus griseus*
Pilot whale	*Globicephala macrorhynchus*
Rough-toothed dolphin	*Steno bredanensis*
Melon-headed whale	*Peponocephala electra*
Fraser's dolphin	*Lagenodelphis hosei*
Pygmy killer whale	*Feresa attenuata*
False killer whale	*Pseudorca crassidens*
Killer whale	*Orcinus orca*
Dall's porpoise	*Phocoenoides dalli*
Whales	
Rorquals	
Blue	*Balaenoptera musculus*
Minke	*B. acutorostrata*
Sei	*B. borealis*
Bryde's	*B. edeni*
Ziphiids (beaked)	*Mesoplodon* spp.; *Ziphius cavirostris*
Sperm	*Physeter macrocephalus*
Dwarf/pygmy sperm	*Kogia* spp.

Appendix 5.2. Relationship between dolphin school size (numbers) and the catch (in short tons) of associated yellowfin tuna. Data collected by SWFC observers abroad U.S. purse seiners, 1981 through 1985.

| | COAST–110°W LONGITUDE | | | | | | 110°W–140°W LONGITUDE | | | | | |
| | Porpoise fish[a] | | School fish[b] | | All | | Porpoise fish[a] | | School fish[b] | | All | |
YEAR	n	(%)	n	(%)	n	(%)	n	(%)	n	(%)	n	(%)
5°N–25°N Latitude												
1981	374	(97.6)	102	(79.4)	476	(93.7)	328	(95.7)	78	(24.4)	406	(82.0)
1982	378	(98.4)	67	(80.6)	445	(95.7)	179	(96.6)	74	(33.8)	253	(78.3)
1984	173	(95.4)	1	(–)	174	(95.4)	153	(94.8)	2	(–)	155	(93.5)
1985	728	(93.0)	24	(–)	752	(91.8)	157	(99.4)	0	(–)	157	(99.4)
1981–85	1,653	(95.5)	194	(76.8)	1,847	(93.6)	817	(96.4)	154	(28.6)	971	(85.7)
5°N–15°S Latitude												
1981	44	(75.0)	122	(33.6)	166	(44.6)	43	(81.4)	0	(–)	43	(81.4)
1982	166	(87.4)	55	(38.2)	221	(75.1)	68	(83.8)	4	(–)	72	(83.3)
1984	27	(92.6)	0	(–)	27	(92.6)	1	(–)	0	(–)	1	(–)
1985	11	(–)	0	(–)	11	(–)	0	(–)	0	(–)	0	(–)
1981–85	248	(86.3)	177	(35.0)	425	(64.8)	112	(83.0)	4	(–)	116	(82.8)

NOTES: These 1981–85 data are of sets (= purse seine launches) during a period straddling the 1982–83 El Niño event and during which time fishermen have often checked tuna schools for size before setting on the school. The data were collected by SWFC technicians aboard U.S. purse seiners. % = percentage of *n* tuna school set upon by the seiners that were with birds. (–) indicates unreliable percentage because of small sample size. There is no data for 1983.

[a] "Porpoise fish" = tuna schools associated with dolphins.

[b] "School fish" = not associated with dolphins.

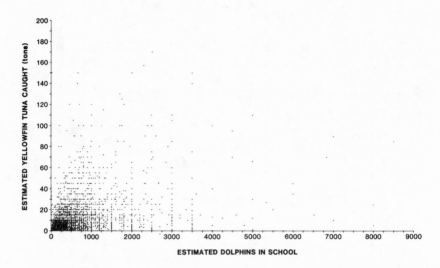

Appendix 5.3. The percent of yellowfin tuna schools (n) (set upon by purse seiners) that were with birds, according to area and whether schools were with dolphins ("porpoise fish") or not ("school fish").

6

Interactions Between Gulls and Otariid Pinnipeds: Competition, Commensalism, and Cooperation

Raymond Pierotti • *Department of Zoology*
University of Wisconsin, Madison

■

Gull Otariid Interactions: Competition or Cooperation?

Gulls and otariid pinnipeds (sea lions and fur seals) exist sympatrically in a large number of marine environments. These include the colder waters of the Pacific basin, including the California and Humboldt Current systems, the northern Kuroshio, the Bering Sea, south Australia, and New Zealand. In the Atlantic basin, these groups occur sympatrically in the Benguela Current system off southwest Africa and along the southeastern coast of South America. Gulls and fur seals of the genus *Arctocephalus* also exist sympatrically on most subantarctic islands. Gulls are also common in a number of areas where otariids are absent, most notably the north Atlantic and the Arctic Ocean. Both groups show an antitropical distribution, with concentrations occurring in or near zones of upwelling and high marine productivity (Davies 1958; Lipps and Mitchell 1976).

It has been suggested that marine mammals and birds compete for limited food resources (Furness 1984), and that reductions in marine mammal populations may have been at least partially responsible for the dramatic increase in gull numbers during the period from 1920 to 1975 (Pierotti 1980). The historical size of gull populations is unknown, but it is known that gulls reached very low population sizes during the nineteenth century (Graham 1975). Gull (and other seabird) population sizes appeared to reach an asymptote in the late 1960s or early 1970s, and declined slightly during the 1970s and 1980s when marine mammal populations have grown rapidly (Furness and Monaghan 1986; Beddington, Beverton, and Lavigne 1985).

Otariid pinnipeds depend on isolated insular rookeries to give birth (Bartholomew 1970), and many large gull colonies are located on islands (e.g., Vermeer 1963; Schreiber 1970; Pierotti 1981, 1982). Thus, there is a possibility that competition exists for breeding space between otariids and gulls in some areas (Warheit and Lindberg, this volume).

There is also a possibility that mutualism or commensalism exists between gulls and otariids. Gulls, which are known to act as scavengers (Vermeer 1963; Schreiber 1970; Spaans 1971; Hunt 1972; Pierotti and Annett 1987), may remove large amounts of carrion (including placentas and dead pups) and waste material (including feces and regurgitated material) from pinniped rookeries. In addition, gulls produce distinctive warning vocalizations in the presence of predators and might thereby act as sentinels for otariid agregations, much as cattle egrets or tickbirds do with ungulate aggregations.

Gulls are commonly found associating with feeding marine mammals of various species, including otariids (Ryder 1957; Sealy 1973; Hoffman, Heinemann, and Wiens 1981; Bayer 1983; Duffy 1983; Pierotti personal observations). This creates the possibility that gulls and otariids may forage cooperatively on schooling marine fishes and invertebrates.

In this essay I review interactions between gulls and otariids, including foraging and breeding distributions (both temporal and spatial), breeding habitat preferences, diets, and direct observations of interactions. I present data on the reproductive success of gulls breeding both in the presence and absence of sea lions. I also compare interactions between gulls and otariids with those between gulls and phocid ("true," earless) seals.

■ Distribution

Otariids are presumed to have evolved along the western coast of North America during the early Miocene (Repenning, Ray, and

Grigorescu 1979). The northern Pacific basin still exhibits the greatest diversity of otariids with four resident species (two endemic fur seals, two endemic sea lions). From this area sea lions probably radiated during the Pliocene along the eastern rim of the Pacific. Sea lions initially invaded the west coast of South America (endemic sea lion), with this species eventually invading the Atlantic as far north as Uruguay and southern Brazil (Vaz-Ferreira 1975a; Repenning, Ray, and Grigorescu 1979). Subsequently, sea lions invaded Australia (endemic sea lion) and New Zealand (endemic sea lion).

Fur seals (subfamily Arctocephalinae) also moved south along the eastern rim of the Pacific during the late Pliocene, leaving endemic species along the Pacific coast of California and Baja California, the Galapagos, and the South American mainland and its offshore islands (Repenning, Peterson, and Hubbs 1971; Repenning, Ray, and Grigorescu 1979). From there fur seals became distributed around the Southern Hemisphere, colonizing all subantarctic islands, South Africa, Australia, and New Zealand. Today, otariids can be found in all marine environments where water temperatures are less than 20°C throughout the year (Davies 1958; Lipps and Mitchell 1976; Repenning, Ray, and Grigorescu 1979). Fur seals tend to inhabit more extreme latitudes than sea lions, which are restricted to temperate waters (King 1964, 1983).

Gulls are found predominantly in the Northern Hemisphere in both the Atlantic and Pacific oceans. Like otariids, gulls exhibit a largely antitropical distribution, being relatively uncommon in the marine tropics (Harrison 1983). Relatively few species of gulls (fourteen) have distributions predominantly or exclusively in the Southern Hemisphere, compared with thirty-two species in the Northern Hemisphere (Harrison 1983).

In those areas where gulls and otariids co-occur, there are more gull species where there are more otariid species (table 6.1): 1) northern Pacific basin: four otariids and eight gulls (including two kittiwakes); 2) Pacific South America: two otariids and four gulls; 3) Australia: two otariids and three gulls; 4) Galapagos and New Zealand: two otariids and two gulls; 5) South Africa: one fur seal and two gulls; and 6) many subantarctic islands: one fur seal and one gull.

Outside of the breeding season when many gulls migrate into marine temperate zones this pattern becomes even more dramatic. The temperate north Pacific may have fourteen to fifteen species of gulls that forage sympatrically in coastal waters, and the ranges of

Table 6.1. Timing of reproduction of otariids and gulls that overlap in distribution during the breeding season.

Species of Otariid	Time of Reproductive Events	Sympatric Gull Species	Time of Reproductive Events
North Pacific			
Steller's sea lion (*Eumetopias jubatus*)	Pups born: late May–early July[a]	Glaucous-winged gull[b] (*Larus glaucescens*)	Eggs laid: May–June
	Lactation: through Aug.		Chicks reared: June–July
		Western gull[c] (*L. occidentalis*)	Eggs: late April–June
			Chicks: June–Aug.
		Glaucous gull[d] (*L. hyperboreus*)	Eggs: May–June
			Chicks: June–Aug.
		Slaty-backed gull[d] (*L. schistisagus*)	Eggs: May–June
			Chicks: June–Aug.
		Black-legged kittiwake[d] (*Rissa tridactyla*)	Eggs: May–July
			Chicks: June–Aug.
		Red-legged kittiwake[d] (*Rissa brevirostris*)	Eggs: June–July
			Chicks: July–Aug.
Northern fur seal (*Callorhinus ursinus*)	Pups born: June–July[e]	Same six species as Steller's sea lion	
	Lactation: through Sept.		
Guadalupe fur seal (*Arctocephalus townsendi*)	Pups born: May–June[f]	Western gull only	Eggs: April–July
	Lactation: through Oct.		Chicks: May–Aug.
California sea lion (*Zalophus californianus*)	Pups born: May–June[a]	Western gull[c]	Eggs: April–June
	Lactation: through Nov.		Chicks: May–Aug.

Galapagos Islands

California sea lion
(Z. c. wollbaeki)
Pups born: Oct.–Dec.[e]
Lactation: through April

Heerman's gull[d]
(L. heermani)
Eggs: April–May
Chicks: May–July

Galapagos fur seal
(A. galapagoensis)
Pups born: Nov.–Dec.[e]
Lactation: through April

Yellow-legged gull[d]
(L. livens)
Eggs: April–May
Chicks: May–July

Lava gull[d]
(L. fuliginosus)
Eggs: Sept.–Nov.
Chicks: Oct.–Jan.

Swallow-tailed gull[d]
(Creagus furcatus)
Eggs: Sept.–Nov.
Chicks: Oct.–Dec.

Southern South America

South American sea lion
(Otaria byronia)
Pups born: Dec.–Jan.[a]
Lactation: through June

Dolphin gull[d]
(L. magellanicus)
Eggs: Dec.–Jan.
Chicks: Jan.–March

South American fur seal
(A. australis)
Pups born: Nov.–Dec.[e]
Lactation: through July

Band-tailed gull[d]
(L. belcheri)
Eggs: Oct.–Dec.
Chicks: Nov.–Jan.

Kelp gull[d]
(L. dominicanus)
Eggs: Sept.–Dec.
Chicks: Oct.–Jan.

Brown-hooded gull[d]
(L. maculipennis)
Eggs: Oct.–Nov.
Chicks: Nov.–Dec.

South Australia

Australian sea lion
(Neophoca cinerea)
Pups born: Dec.[a]
Lactation: through June

Pacific gull[d]
(L. pacificus)
Eggs: Sept.–Nov.
Chicks: Oct.–Jan.

Australian fur seal
(A. forsteri)
Pups born: Dec.[e]
Lactation: through May

Silver gull[d]
(L. novaehollandiae)
Eggs: Nov.–Dec.
Chicks: Dec.–Feb.

Table 6.1. (Continued)

Species of Otariid	Time of Reproductive Events	Sympatric Gull Species	Time of Reproductive Events
New Zealand			
New Zealand sea lion (*Phocarctos hookeri*)	Pups born: Dec.–Jan.[a] Lactation: through April	Red-billed gull[d] (*L. novaehollandiae*)	Eggs: Oct.–Nov. Chicks: Nov.–Jan.
New Zealand fur seal (*A. forsteri*)	Pups born: Dec.–Jan.[a] Lactation: through July	Kelp gull[d]	Eggs: Oct.–Nov. Chicks: Nov.–Jan.
South Africa			
South African fur seal (*A. pusillus*)	Pups born: Nov.–Dec.[c] Lactation: through the next Oct.	Kelp gull[d]	Eggs: Sept.–Dec. Chicks: Oct.–Feb.
Subantarctic Islands			
Subantarctic fur seal (*A. tropicalis*)	Pups born: Nov.–Dec.[e] Lactation: through April	Kelp gull[d]	Eggs: Oct.–Dec. Chicks: Nov.–Feb.
Antarctic fur seal (*A. gazella*)	Pups born: Dec.–Jan.[e] Lactation: through April		

[a] King 1964, 1983.
[b] Vermeer 1963; Ward 1973.
[c] Pierotti 1981.
[d] Harrison 1983.
[e] Gentry and Kooyman 1986.
[f] Pierson 1978.

the four resident otariids overlap to a much greater extent during winter than they do during the breeding season (Bartholomew 1970; author's unpublished observations). In the Southern Hemisphere gull species diversities, even in productive areas, i.e., the Humboldt Current, are lower for several reasons. First, as mentioned above, there are fewer species of gull in total. Second, the Antarctic mainland has only one resident gull, compared with eight in the Arctic basin that migrate into temperate waters. Finally, there are relatively few large temperate bodies of land with fresh water where gulls nest and then migrate to coasts outside of the breeding season (thirteen species in the Northern Hemisphere employ this pattern compared with six in the Southern Hemisphere; Harrison 1983).

The distinguishing feature of areas where otariids and gulls occur sympatrically is the occurrence of strong seasonal upwelling which is especially pronounced in eastern boundary currents (Wooster and Reid 1963; Richards 1981). Upwelling zones typically have high standing crops of fish and squid (MacCall 1984). The highest diversity of otariids and gulls is found along the west coast of North America where the California Current is a major zone of upwelling, and productivity of zooplankton is high from January through August (Bolin and Abbott 1963; MacCall 1984). The next most diverse assemblage of otariids and gulls is found in the highly productive Humboldt Current system, which is more variable in productivity than the California Current because of the periodic El Niño oscillations (Quinn et al. 1978). The Benguela Current off southwest Africa, another eastern boundary current, also has gulls and fur seals. New Zealand and the southern coast of Australia are also zones of upwelling, as is the Antarctic Convergence.

■

Patterns of Habitat Use

Location of Breeding Colonies

Within zones of upwelling and high marine productivity otariids are much more restricted in their breeding colonies than are gulls. Otariids must give birth on land and during the same season be able to feed in the marine environment. In addition, adaptations that allow thermoregulation and the pursuit of fast-moving prey in the marine environment, i.e., large, bulky bodies and limbs reduced to flippers, restrict mobility on land and make all pinnipeds vulnerable to terrestrial predators whenever they are out of the water (Bartholomew 1970). This combination of factors restricts pinnipeds to breeding or hauling out either on islands or on beaches or caves at the foot of cliffs that are so steep that terrestrial predators cannot climb them. On large islands pinnipeds are typically found only around the periphery because their limited terrestrial mobility restricts them to beaches or rocky marine terraces with shallow slopes where they can enter and leave the water with relative ease (Peterson and Bartholomew 1967).

Gulls nest in a variety of habitats, including small offshore stacks, harbor markers, offshore rocks accessible from the mainland, and even roofs of buildings (Tinbergen 1960). Large gull colonies are, however, restricted to islands because large colonies on the mainland (except those in marshes) attract predators. Islands used by gulls are often smaller and have much steeper slopes than those used by otariids.

As an example, I compare sea lion and gull distributions along the west coast of North America where the pattern is best documented. In California the largest breeding colony of gulls *(Larus occidentalis)* is on Southeast Farallon Island (15,000 pairs; Sowls et al. 1980). Southeast Farallon is a marine seamount with steep slopes except on its southern exposure where there is an extensive gently sloping marine terrace. At present relatively few otariids breed on the island, with a small population of Steller's sea lions *(Eumetopias jubatus)* (approximately 100 females and 15–20 males), and at least in the mid-1970s, one female California sea lion *(Zalophus californianus)* that gave birth, and about 300–400 adult male *Zalophus* (Pierotti et al. 1977). These otariids are concentrated on small beaches and marine terraces on the sheltered north side of the island, and on the west end of Saddle Rock, a gently sloping offshore rock (author's observations).

Western gulls nest in all habitats above the splash zone on Southeast Farallon, with the exception of steep cliffs, and areas dominated by Brandt's cormorants *(Phalacrocorax penicillatus)* and common murres *(Uria aalge)* which are able to competitively displace the gulls (author's observations). Gulls nest at high densities over the marine terrace (typical internest distances are 2–3m). It is unlikely that this was the case two hundred years ago, however, since Southeast Farallon may have been the largest breeding colony of Guadalupe fur seals *(Arctocephalus townsendi)*. During the period between 1800 and 1810 Russian and American sealers took as many as 150,000 fur seals off Southeast Farallon (Busch 1985). This total included northern fur seals *(Callorhinus ursinus)*, and possibly California and Steller's sea lions as well. Regardless of the species composition, however, the only area on Southeast Farallon that could have accommodated the 30,000–40,000 pinnipeds necessary to sustain such a harvest is the southern marine terrace (Ainley and Lewis 1974; author's observations).

As a result, this area was probably unavailable for gull nesting until the nineteenth century. This may have been a moot point, however, since the gulls themselves were exploited heavily by eggers during much of the nineteenth and early twentieth centuries (Ainley and Lewis 1974). Therefore gull numbers were probably much lower than at present, and there was probably sufficient habitat for gulls on the steeper central, northern, and western portions of the island.

Other large western gull colonies in California show somewhat different patterns (data on gull population sizes are from Sowls et al. 1980; data on pinnipeds are from Bartholomew 1967 and author's data). Middle Anacapa Island (2,500 pairs) has very steep cliffs around the periphery and is only used by sea lions for hauling out. Santa Barbara Island (1,200 pairs) also has steep cliffs around its periphery, but there are cobble beaches used by *Zalophus* for breeding around much of the periphery. There is, however, no competition for space between gulls and sea lions on this island since gulls nest in vegetated areas atop the island where sea lions never climb. San Nicolas Island (900 pairs) is a large island (21,000 acres) and has large breeding populations of *Zalophus* (5,000–7,000; Peterson and Bartholomew 1967), but as on Santa Barbara Island, sea lions are on beaches below the cliffs at the water's edge and the gulls nest on hillsides inland from the cliff edge (Schreiber 1970). Castle Rock (675 pairs) is a steep offshore rock whose periphery is used as nonbreeding haulout by *Zalophus* and *Eumetopias*. Prince Island (480 pairs) is a steep rock off the shore of San Miguel Island, which is one of the main breeding grounds for *Zalophus* and also has (or recently had) small breeding populations of *Eumetopias* and *Callorhinus*.

There are some islands, however, where pinniped use patterns may reduce or prevent gull nesting. Año Neuvo Island is a small, low, sandy island with a few marine terraces on its seaward exposure. It is used as a breeding area by western gulls (150 pairs) and *Eumetopias* (800 females and 100 males in 1972), and as a haulout by a varying number of nonbreeding male *Zalophus* (counts during gull breeding season ranged from 1,000–6,000 individuals; author's data). *Eumetopias* breed on the marine terraces and on offshore rocks, and *Zalophus* primarily use the beaches and the sandy interior of the island. This use pattern appears to restrict gull breeding to several small rock outcroppings and to a vegetated area on the north end of the island. Some gulls nest in flat sandy areas, but eight of seventeen nests were destroyed by male *Zalophus,* and six of the remaining nine nests had crushed or broken eggs, presumably from sea lion disturbance (author's observations).

In Oregon, Washington, British Columbia, and Alaska gulls nest primarily on ledges or the tops of islands with steep cliffs, and sea lions and fur seals use caves or beaches at the foot of the cliffs. North of San Francisco, there are numerous islands suitable for gulls (Sowls et al. 1980), whereas sea lions only breed in a few locations (Mate 1975). Those areas used by otariids for hauling out probably have little or no impact on gull breeding populations.

Similarly, in other areas of the world where otariids and gulls breed sympatrically, otariids appear to congregate on beaches or terraces at or near the edge of the water (Gentry and Kooyman 1986). Since most breeding islands are large and often have steep interiors, e.g., Guadalupe, the Falklands, the Galapagos, and the subantarctic islands, there seems to be abundant nesting space for gulls, and other seabirds, that is not used by otariids. In summary, it appears that otariids and other pinnipeds do not compete with gulls for breeding space, except on small islands with large flat areas, and even in these locations it is doubtful that sea lions have a major negative impact on gull populations.

Temporal Patterns of Habitat Use

In general, there is extensive temporal overlap between the breeding seasons of gulls and otariids (table 6.1). Typically gulls establish nesting territories in early spring in both the Northern and Southern Hemispheres and remain on those territories until chicks are independent in midsummer (Tinbergen 1960; Pierotti 1981, 1982). This depends to a large degree on the climate around the breeding colony. In mild climates, e.g., California, gulls may establish territories in January–February and remain until late August (Pierotti 1981), whereas in Alaska, territory establishment is delayed until May, and birds may remain on the territory until September (E. Murphy, personal communication).

In the Southern Hemisphere, Australian gulls establish territories in August–September and remain until March, whereas subantarctic gulls establish in November and depart in February or March (Harrison 1983; Corxall 1984). These dates are determined by migration patterns, the availability of food around breeding colonies, and the presence of predators (gulls establish territories at later dates and do not initiate incubation until the clutch is complete on colonies where terrestrial predators may occur; Tinbergen 1960).

In otariids, an analogous situation occurs. In the Northern Hemisphere adult males of both fur seals and sea lions typically arrive on traditional breeding grounds before females, and set up territories in mid to late spring (May to early June). Females arrive throughout June and early July and give birth a few days after arrival (King 1983). In high latitudes, e.g., Alaska, rookeries are typically abandoned in early autumn, whereas lower-latitude colonies, e.g., California's Channel Islands, rookeries may be occupied year round (Gentry and Kooyman 1986).

In the Southern Hemisphere, adult male fur seals typically establish territories in October, and the females arrive and give birth in November and December (Gentry and Kooyman 1986). Sea lions in the Southern Hemisphere initiate territory establishment and give birth about a month after fur seals (King 1983). At higher latitudes both groups abandon the rookeries in March–April, whereas in lower latitudes, i.e., the Galapagos and Namibia, colonies may be occupied throughout the year (King 1983; Vaz-Ferreira 1975a, 1975b; Gentry and Kooyman 1986).

This overlap in timing of breeding between gulls and otariids increases the potential for competition for both breeding space and food. Indeed, young, inexperienced gulls attempt to establish territories in areas used by otariids for breeding before the otariids arrive (author's observations). These individuals are invariably displaced, either as a result of having their nests destroyed by bull otariids during territory establishment or by swamping and nest destruction when females arrive. As discussed above, however, it is unlikely that the prevention or delay of breeding by these few pairs has much of a negative impact on gull populations.

In areas where gulls and otariids breed sympatrically there is considerable overlap in diet (table 6.2). Therefore, potential competition exists for food. There are, however, alternative explanations for the overlap in diet. First, it is likely that the timing of breeding in both gulls and otariids are adjusted to similar phenomena. As mentioned above, these groups inhabit areas of strong upwelling, and June and July are the period when abundances of fish and squid in upwelling zones are at their peaks. Many species of fish spawn

Table 6.2. Comparison of the diets of Western gulls and pinnipeds in California with abundance data from fisheries.

			RANK IN ABUNDANCE IN DIET			
Rank in Fishery[a]	Western Gull[b]	California Sea Lion[c]	Steller's Sea Lion[d]	Northern Fur Seal[a]	Harbor Seal[d]
1 Northern anchovy (*Engraulis mordax*)	Northern anchovy	Pacific whiting	Rockfish	Northern anchovy	Surfperch (Embiotocidae)
2 Pacific whiting (*Merluccius productus*)	Rockfish (juv.)	Rockfish (juv.)	Pacific whiting	Pacific whiting	Eelpout (Zoarcidae)
3 Rockfish (*Sebastes* spp.)	Plainfin midshipman (*Porichthys notatus*)	Northern Anchovy	Spotted cusk eel	Market squid	Flatfish (Pleuronectidae)
4 Jack mackerel (*Trachurus symmetricus*)	Pacific Whiting	Plainfin midshipman	Flatfish	Pacific saury	Greenling (Hexagrammidae)
5 Pacific saury (*Cololabis saira*)	Jack mackerel	Market squid	Sablefish (*Anopoploma*)	Rockfish	Sculpin (Cottidae)
6 Market squid (*Loligo opalescens*)	Market squid	Spotted cusk eel	Lingcod (*Ophiodon*)	Jack mackerel	Octopus (*Octopus* spp.)
	White croaker (*Genyonemus lineatus*)	White croaker	Jack mackerel		Pacific herring

			Lingcod
Spotted cusk eel (*Chilara taylori*)	Jack mackerel	Plainfin midshipman	
Pacific herring (*Clupea pallasii*)	Pacific herring	Northern anchovy	
Pacific tomcod (*Microgadus proximus*)	Pacific tomcod	Market squid	

[a]Kajimura 1985. [b]Hunt and Hunt 1975; Pierotti 1981; author's unpublished data. [c]Antonelis et al. 1984. [d]Jones 1981.

when zooplankton abundance peaks in May and June, and other species are in the area to feed on zooplankton or small fish. This abundance continues as larger species, e.g., cod, hake, and squid, follow prey species, e.g., capelin or anchovy, into these areas (MacCall 1984). Therefore, the overlap in diet may be coincidental, with both groups exploiting similar abundant prey species (Furness 1984), or it may result, at least in part, from cooperative foraging between gulls and otariids (see below).

■

Interactive and Cooperative Behavior Between Gulls and Otariids

Gulls as Scavengers

Gulls are opportunistic feeders that exploit a wide range of food types, including scavenging on garbage dumps and around fish processing operations (Vermeer 1963; Pierotti and Annett 1987). Gulls around otariid rookeries act as scavengers on carcasses, expelled placentas, fecal material, and regurgitated food (Schreiber 1970; Hunt and Hunt 1975; author's observations). It has been suggested that sea lion placentas may represent an important food source to gulls (Schreiber 1970; Hunt and Hunt 1975). Since 5–15 percent of otariid pups typically die before leaving the rookery, and more than 70 percent may die during poor years (Peterson and Bartholomew 1967; Gentry and Kooyman 1986), there are large amounts of carrion available for gulls. Since one of the most common causes of death in newborn otariid pups is disease (King 1964, 1983), it is possible that gulls benefit otariids when they remove from rookeries carrion, placentas, and fecal material that could act as breeding areas for disease.

Gulls as Sentries

It is also likely that otariids benefit from gulls acting to warn them about the presence of potential predators. Gulls have two basic alarm vocalizations, a short "hah-hah-hah" which is of low intensity and a sharp, high-pitched "kew" which is of high intensity (Tinbergen 1960). Sea lions are very attentive to gull alarm calls, and appear to respond to the two calls in much the same manner as do gulls. In 97 percent of the situations where a gull on a sea lion rookery or flying overhead gave the "hah-hah" call (n = 108), sleeping sea lions raised their heads and looked around, and active sea lions looked intently in the direction of the gull. When gulls gave the "kew" call (n = 65), all sea lions rose up and milled around rapidly, with some individuals moving into the water. On six occasions when the sea lions saw a potential predator (an investigator), they all dashed rapidly to the water's edge and dove into the water. Some territorial bulls did not flee, and young pups (less than one week old) also remained.

This response on the part of sea lions is not simply a response to loud noise. Sea lions ignored 85 percent of gull long calls (n = 126) and all loud vocalizations by oystercatchers (n = 34), and appeared to understand the significance of the gull vocalizations. This is an appropriate learned response on the part of sea lions, since for much of their shared evolutionary history the two groups have had a number of predators in common, including foxes, coyotes, eagles (on otariid pups), wolves, bears, wolverines, *Homo sapiens*, and hyenas (on South African fur seals).

Mixed-Species Foraging Aggregations

Otariids and gulls have often been observed together in foraging aggregations. On all thirty-seven occasions that I have observed sea lions feeding at sea they were accompanied by large numbers of gulls. These included thirteen observations with *Zalophus* as the only pinniped present, twelve observations with *Eumetopias* as the only pinniped present, and twelve observations with both species feeding together in mixed groups. These sea lions were always accompanied by western gulls. Other species of seabird were present on various occasions (including Bonaparte's, Heerman's, and glaucous-winged gulls, black-legged kittiwakes, pelicans, cormorants, terns, and alcids). During surveys of marine bird and mammal abundance in northern and southern California, on every occasion when sea lions were observed foraging they were accompanied by western gulls and other species of gulls (K. T. Briggs, pers. comm.). Other authors have reported similar observations in Alaska (Hoffman, Heinemann, and Wiens 1981), Oregon (Bayer 1983), and California (Ryder 1957; Baltz and Morejohn 1977), and off Peru (Duffy 1983).

On seven occasions, I was able to follow the development of a foraging aggregation involving western gulls and sea lions from its earliest stages. In each case, western gulls appeared to play an active role in the formation of the aggregation. Typically western gulls were either the first bird to spot a prey school and their circling behavior above the school attracted both other marine birds and marine mammals, or their movement toward small groups of birds or mammals working a prey school attracted large numbers of other species to the area.

For example, on June 19, 1975, a stream of western gulls was observed to depart Santa Barbara Island and head in the direction of a group of gulls wheeling over a commotion on the surface about 3 kilometers offshore. A group of female *Zalophus* milling in the surf swam out to sea, moving as a group in the same line as the gulls. The sea lions could not have seen the wheeling gulls from the base of the cliff. As the line of gulls increased, low-flying brown pelicans and Brandt's cormorants flying across their path changed direction to follow the flight path of the gulls (see also Hoffman, Heinemann, and Wiens 1981 for illustrations of similar phenomena). Within one half-hour, a feeding aggregation had developed, which lasted for approximately 35 minutes. Gulls returning from this group regurgitated northern anchovies *(Engraulis mordax)*. On July 18, 1975, I observed the formation of a similar aggregation in the same manner, but no pelicans were observed. The prey species in this case was Pacific Saury *(Cololabis saira)*.

Similarly, from the highest point on Southeast Farallon Island I observed western gulls flying out to an aggregation of Brandt's cormorants diving as a group about 1 kilometer offshore. After a group of gulls had gathered and were circling overhead and dividing to the surface both Steller's and California sea lions headed toward the aggregation. Once the sea lions reached the area and began diving along with the cormorants and gulls, many more gulls joined the aggregation.

An alternative hypothesis that could explain the observation that sea lions joined foraging aggregations from a distance is that sea lions produce underwater vocalizations that attract other sea lions. This is unlikely for three reasons. First, sea lions join aggregations from considerable distances, i.e., 2 or more kilometers. Second, sea lions have been observed to join groups where no sea lions were present. Finally, hydrophones placed underwater near sea lions foraging in groups revealed that sea lions did not vocalize under these conditions (author's data).

All thirty-seven foraging aggregations involving sea lions and gulls that I observed had substantially the same structure. Sea lions dove and surfaced as tightly coordinated groups of 40–50 (the complete range was from 10 to more than 100). Groups of hundreds to thousands of sea lions have been observed traveling to feeding grounds in Alaska, where they broke into smaller feeding groups of less than 50 after arrival (Fiscus and Baines 1966).

Sea lion groups I observed swam underneath schools of fish (primarily northern anchovy) and then drove the fish to the surface where the fish concentrated in tightly milling balls. When these balls of fish were at the surface gulls would dive into them and pick fish off the surface. At the same time, the sea lions surfaced through the fish which sent fish flying in all directions. The surface of the water was a milling mass of gulls and sea lions taking fish. On some occasions cormorants, loons, or murres were also be part of this activity, but most diving birds avoided these areas of intense activity (see also Hoffman, Heinemann, and Wiens 1981 for similar observations).

Otariids often feed as individuals and dive deep for prey (Gentry and Kooyman 1986), but they appear to forage most efficiently during shallow dives in groups (Rand 1959; Fiscus and Baines 1966; Gentry and Kooyman 1986). During my observations individuals or groups of two sea lions would try to take fish, but the schooling behavior of the fish enabled the fish to confuse and escape from these small groups (C. Annett, personal communication). Group foraging is much more efficient for sea lions in exploiting schooling fishes since groups of three or more showed a marked increase in capture rate (Table 6.3).

Based on extensive data from diving female otariids, a series of "rules" have been developed for efficient foraging in otariids (Gentry and Kooyman 1986; 253–256). First, dive no deeper or longer than

Table 6.3. Capture success rate of California sea lions feeding in groups on northern anchovies.

Number of Sea Lions in Group	Number of Attempted Captures	Number of Successful Captures	Percentage of Successful Captures
1	63	2	3.2
2	82	43	52.4
3	47	38	80.9
4 or more	132	113	85.6

necessary; ideally, search and feeding should occur at the surface. Second, search during the descent, and descend as quickly as possible. Third, search for, pursue, and capture prey on the ascent because this silhouettes prey against the bright surface (Hobson 1966). Finally capture only large or energy-rich prey on deep dives.

The picture that emerges from these rules and from observations of otariid foraging groups (see above and table 6.3) suggests that otariids forage most efficiently when they 1) locate large concentrations of prey near the surface, 2) can dive shallowly underneath the school of prey and force it to the surface, which 3) concentrates the prey into balls where they can be easily seen and captured (Grover and Olla 1983; author's observations).

Gulls have a wide variety of foraging techniques, but a very limited ability to plunge dive and cannot pursue prey underwater (Ashmole 1971). As a result, to take schooling fish gulls require the presence of diving birds or marine mammals to drive prey to the surface. Gulls have been observed taking fish driven to the surface by large predatory fishes (Coblentz 1985; author's data), pinnipeds, cetaceans (Wursig and Wursig 1980; Pierotti, this volume) and diving seabirds (Bartholomew 1942; Hoffman, Heinemann, and Wiens 1981; Grover and Olla 1983).

In order to facilitate foraging at sea, natural selection may have acted on gulls to increase individual foraging efficiency through the formation or maintenance of pelagic foraging aggregations. Gulls forage more efficiently as part of a group than as individuals (Gotemark, Winkler, and Anderson 1986), so the original selective pressure may have acted through the formation of conspecific groups. Investigators studying the structure of mixed-species seabird foraging aggregations, however, have characterized gulls (including kittiwakes) as "catalysts" (Sealy 1973; Hoffman, Heinemann, and Wiens 1981). Catalysts are defined as "birds whose foraging and feeding behaviors are highly conspicuous ... the arrival of a catalyst was necessary for flock development" (Hoffman, Heinemann, and Wiens 1981).

If other species of seabird recognize gulls as indicators of food it would be surprising if marine mammals also did not learn to use this information. Therefore, both gulls and otariids would benefit as individuals from the association when foraging. Sea lions are able to move around as individuals or small groups to search for widely scattered, patchily distributed food sources and can coalesce quickly when prey schools are located by using gulls as indicators of activity. Gulls obtain access to schools of prey driven to the surface that would otherwise be difficult or impossible for them to exploit. In addition, sea lions are better at concentrating prey and driving them to the surface than are diving birds. The rate of prey capture success for western gulls was higher when sea lions were present than when gulls were feeding only with other seabirds (table 6.4).

This suggests that despite the potential for competition that exists between otariids and gulls, the most important component of the interaction between these groups may be cooperative foraging. I do not mean to suggest that all, or even most, foraging by gulls and otariids occurs in aggregations containing both of these groups. Gulls often forage either alone, or with other seabirds or marine mammals (Hoffman, Heinemann, and Wiens 1981; author's observations), and most foraging dives by otariids appear to go to depths of 30–50 meters or more (Gentry and Kooyman 1986). If the foraging efficiency of both gulls and otariids is increased by the association, however, it would be favored by selection and could be an important component of foraging activities in these groups.

Some support for this idea is provided by examining the data on the reproductive output of gulls breeding on colonies where sea lions outnumbered the gulls present compared with colonies with either small populations of sea lions or no sea lions (table 6.5). Southeast Farallon has a small population of sea lions relative to the number of gulls and Moss Landing has no sea lions within 30 kilometers of the colony. Both gull colonies have abundant food supplies (Pierotti 1981; Pierotti and Bellrose 1986) and the gulls

---

Table 6.4. Rates of successful prey capture of western gulls foraging in assemblages with sea lions and with other Seabirds.

ASSEMBLAGE WITH SEA LIONS			ASSEMBLAGE WITH SEABIRDS ONLY		
Number of attempted captures	Number of successful captures	Percentage successful	Number of attempted captures	Number of successful captures	Percentage successful
231	147	63.6	192	81	42.1
146	121	82.9	132	41	31.1
303	218	71.9	174	33	19.0
60	42	70.0	224	54	24.1
183	116	63.4	236	62	26.3

Table 6.5. Breeding success of gull colonies with and without large populations of sympatric otariids.

Colony and Year	Mean Clutch Size	Mean Hatch Success	Mean Fledgling Success
Western gull (w/o otariids)			
Southeast Farallon			
1971[a]	2.81	2.20	1.91
1972[a]	2.72	2.25	1.91
1973[b]	2.81	2.37 ± 0.82	1.54 ± 0.94
1974[b]	2.89	2.47 ± 0.77	2.25 ± 0.91
Moss Landing			
1981[c]	2.86 ± 0.54	2.40 ± 0.65	1.79 ± 1.12
1982[c]	2.71 ± 0.72	2.01 ± 0.72	1.90 ± 0.94
Alcatraz			
1983[d]	2.90	2.26	1.00
1984[d]	2.74	1.64	1.12
Western gull (w/otariids)			
Santa Barbara Island			
1972[e]	—	2.53	2.15
1975[e]	2.67 ± 0.81	2.67 ± 0.81	2.25 ± 0.95
Año Nuevo Island			
1974[f]	2.64	2.21	2.13
1976[f]	2.57	1.98	1.96
Glaucous-winged gull			
Mandarte Island			
(w/o otariids)			
1962[g]	2.82 ± 0.02	2.34	1.63 (70%)
1969[h]			(70%)
Cleland Island			
1970[h]			(80%)
Queen Charlotte Island			
(w/otariids)			
1970[h]			(90%)

[a] Data from Coulter 1973; cited in Pierotti 1981.
[b] Data from Pierotti 1981.
[c] R. Pierotti, unpublished data.
[d] R. Pierotti and C. Annett, unpublished data.
[e] Data from Hunt and Hunt 1972; cited in Pierotti 1981.
[f] Data from Briggs 1977.
[g] Data from Vermeer 1963.
[h] Data from Ward 1973.

that nest there produce larger clutches than colonies with large otariid populations. Western gulls on Alcatraz have no sea lions nearby and a poorer food supply. These birds have similar clutch sizes, but low hatching and fledging success compared to other colonies (author's observations and unpublished data).

In contrast, on Año Nuevo and Santa Barbara islands there are large sea lion breeding populations. These islands have western gull colonies that have higher rates of fledging despite smaller clutches (table 6.5). Similarly, glaucous-winged gulls on Mandarte Island have large clutches, but low fledging rates in the absence of sea lions, whereas in the Queen Charlotte Islands and on the outer coast of Vancouver Island there are large breeding populations of *Eumetopias*, and fledging success was much higher (Vermeer 1963, 1982; Ward 1973).

These differences in fledging rates are probably the result of differences in foraging efficiency during the chick-rearing period. Gulls lay their eggs in late April and May before female otariids arrive on breeding colonies (Pierotti 1981; Vermeer 1963). Gulls switch their diets to easily handled but nutritious food around the time that chicks hatch (Tinbergen 1960; Ward 1973; Pierotti and Annett 1987) which is also around the time when female sea lions arrive and give birth (table 6.1). All species of sea lions for which there is data feed on schooling fishes, squids, or other invertebrates during the breeding season (King 1983; Antonelis, Fiscus, and DeLong 1984; Gentry and Kooyman 1986).

For early growth chicks require foods such as fish and reject foods that are more difficult to handle such as intertidal organisms or garbage (Murphy et al. 1984; Pierotti and Annett 1987). Therefore, the gulls that are the most efficient at exploiting fish are more effective at raising young. Since cooperative foraging with sea lions increases the efficiency with which gulls catch fish (table 6.4), gulls that feed with sea lions would be expected to have greater success in raising offspring, regardless of clutch size.

Further support for this idea comes from the observation that western gulls on Año Nuevo and Santa Barbara islands, and glaucous-winged gulls on Cleland and the Queen Charlotte islands feed their young almost exclusively on fish, whereas on Alcatraz and Mandarte Islands gulls feed their young more human refuse (Hunt and Hunt 1975; Ward 1973; Pierotti 1981; Vermeer 1963, 1982; author's observations). Gulls on Southeast Farallon, Alcatraz, and Moss Landing feed fish to their young, but they may be less efficient at capturing fish than gulls from colonies where sea lions are present. Most of their foraging is done either alone or in association with other birds rather than sea lions.

■
Interactions Between Gulls and Phocids

In contrast to the set of interactions observed to occur between gulls and otariids, interations between gulls and phocids appear to be largely antagonistic, and avoidance rather than attraction appears to be the rule between these groups. As with otariids, gulls have been observed to feed on placentas and dead pups of phocids. Phocids breed primarily in mid- to late winter, with the only exceptions to this rule being harbor seals and the British population of gray seals (King 1964; Pierotti and Pierotti 1980). Therefore, most of the scavenging is carried out by juvenile gulls that remain around nesting colonies (Briggs 1977; author's observations). Phocids often threaten and snap at gulls, especially when these birds pick at wounds resulting from shark attacks or breeding activities, or steal the thick (50 percent fat) milk that leaks from the mammary glands of lactating female phocids (Briggs 1977; author's observations). These phenomena have not been observed in gull-otariid interactions, possibly because otariids are mobile enough to discourage such activities by gulls.

Phocids show varying responses to gull warning vocalizations. Large phocids, i.e., elephant seals *(Mirounga)*, were observed to ignore gull alarm calls 92 percent of the time (n = 83). However, the smaller, more dispersed harbor seals *(Phoca vitulina)* always responded by agitatedly looking around (n = 45; author's data). In general, phocids are less responsive than otariids to gull warning calls. Otariids breed at the same time as gulls, and are therefore near attentive breeding adult gulls that emit specific calls under conditions of danger. In contrast, most alarm calls given by gulls outside of the breeding season (367 of 422) are of low intensity and result from squabbles between gulls or from threats directed at gulls by the phocids themselves (author's observations).

There is little overlap in diets between gulls and phocids. In California, western gulls show complete overlap in diet with California sea lions, with the four most common prey species being the same (table 6.2). Western gulls also share eight of the ten most common prey species with Steller's sea lions, and five of six prey species with northern fur seals (table 6.2). In contrast, only one prey species is shared by gulls and harbor seals, and this species (Pacific herring) is taken rarely by both gulls and seals (table 6.2).

Phocids tend to forage solitarily and to take slow-moving solitary benthic fishes and invertebrates (King 1964, 1983; Jones 1981; author's observations). I have observed phocids foraging both during underwater observations and from shore, and have never observed gulls to attend feeding phocids. Another factor that may contribute to this apparent avoidance of phocids by gulls is that some phocids prey on birds. I have observed harbor seals to capture cormorants and murres swimming on the surface, and to grab gulls. Gulls always managed to escape after a struggle. Gulls will not land on the water near swimming *Phoca*, and take off from the surface if approached by phocids (author's observations).

This dramatic difference in response to the two families of pinniped appears to be related to differences in foraging tactics, and benefits that may accrue to gulls from the association. Gulls are attracted to the socially foraging otariids because these pinnipeds drive prey to the surface where they can be exploited by gulls and are not known to prey on gulls. In contrast, gulls receive no benefits from associating with phocids and may in fact be at risk of predation.

■
Summary

The data presented in this paper suggest that, despite the potential for competition for both food and breeding space that exist between gulls and otariids, relations between these species are either neutral or mutually beneficial. The positive relationship between the number of otariid and number of gull species present in biogeographic areas is likely to be coincidental, with both groups showing higher species diversity in areas of high productivity due to upwelling.

The overlap in diet between these groups is probably due to two primary factors. First, both groups exploit the most abundant species of fish and pelagic invertebrates in local areas. However, there remains the question of how gulls, which are poor plunge divers and forage almost exclusively at the surface, obtain access to these food sources, especially schooling fishes and squids. The answer probably lies in the tendency of gulls to associate with diving birds and mammals that to concentrate schooling prey near the surface. On many of the islands where gulls and otariids breed sympatrically, the otariids are the diving species with which the gulls show the greatest foraging efficiency (table 6.4). This association may contribute to the higher rates of chick growth and fledging success observed in these gull colonies (table 6.5).

Only a few colonies that show little or no physical relief, or around the periphery of large colonies, is there likely to be competition for breeding space between gulls and otariids. As a result, it is unlikely that, contrary to suggestions of other authors (Warheit and Lindberg, this volume), gull populations are affected to any degree by this competition. Overall, it is more likely that both gulls and otariids benefit from associations between these groups. Otariids benefit through the removal of carrion from breeding colonies, from antipredator vigilance by gulls, and from more efficient location of patchily distributed schools of prey. Gulls benefit primarily from increased access to schooling prey provided by the foraging activities of otariids that concentrate prey at the surface.

Acknowledgments

I thank Cynthia Annett and Kenneth Briggs for permission to quote unpublished observations. I thank Cynthia Annett, William Drury, and Gerry Sanger for comments on an earlier version of this manuscript, Cliff Fiscus, Robert Jones, and David Sergeant for discussions of pinniped foraging behavior, and John Croxall, D. Duffy, and F. Trillmich for discussion of pinniped-gull interactions.

References

Ainley, D. G., H. R. Huber, and K. M. Bailey. 1982. Population fluctuations of California sea lions and the Pacific whiting fishery off central California. *Fish. Bull.* 80:253–258.

Ainley, D. G. and T. J. Lewis. 1974. The history of the Farallon Island marine bird populations, 1854–1972. *Condor* 76:432–446.

Antonelis, G. A., Jr., C. H. Fiscus, and R. L. DeLong. 1984. Spring and summer prey of California sea lions, *Zalophus californianus*, at San Miguel Island, California, 1978–1979. *Fish. Bull.* 82:67–76.

Ashmole, N. P. 1971. Seabird ecology and the marine environment. In D. S. Farner and J. R. King, eds., *Avian Biology*, 1:223–286. New York: Academic Press.

Bailey, K. M. and D. G. Ainley. 1982. The dynamics of California sea lion predation of Pacific hake. *Fish. Res.* 1:163–176.

Baltz, D. M. and G. V. Morejohn. 1977. Food habits and niche overlap of seabirds wintering on Monterey Bay, California. *Auk* 94:526–543.

Bartholomew, G. A. 1942. The fishing activities of double-crested cormorants on San Francisco Bay. *Condor* 44:13–21.

Bartholomew, G. A. 1967. Seal and sea lion populations of the California islands.

Proceedings of the Symposium on the Biology of the Channel Islands, pp. 229–244. Santa Barbara Natural History Museum Press. Santa Barbara, Calif.

Bartholomew, G. A. 1970. A model for the evolution of pinniped polygyny. *Evolution* 24:546–559.

Bayer, R. D. 1983. Birds associated with California sea lions at Yaquina Estuary, Oregon. *Murrelet* 64:48–51.

Beddington, J. R., R. J. H. Beverton, and D. M. Lavigne. 1985. *Marine Mammals and Fisheries*. London: Allen and Unwin.

Bolin, R. and D. Abbott. 1963. The productivity of the California Current system. California Cooperative Oceanographic Fisheries Investigation Rept. no. 8.

Briggs, K. T. 1977. Social dominance in young western gulls: its importance in survival and dispersal. Ph.D. dissertation, University of California, Santa Cruz.

Busch, B. C. 1985. *The War Against the Seals: A History of the North American Seal Fishery*. Montreal: McGill-Queens University Press.

Coblentz. B. E. 1985. Mutualism between laughing gulls and epipelagic fishes. *Cormorant* 13:61–63.

Croxall, J. P. 1984. Seabirds. In R. M. Laws, ed., *Antarctic Ecology*, pp. 533–618. New York: Academic Press.

Davies, J. L. 1958. The Pinnipedia: an essay in zoogeography. *Geogr. Review* 48:474–493.

Duffy, D. C. 1983. The foraging ecology of Peruvian seabirds. *Auk* 100:800–810.

Fiscus, C. H. and G. A. Baines. 1966. Food and feeding behavior of Steller's and California sea lions. *J. Mammal.* 47:195–200.

Furness, R. W. 1984. Modelling relationships among fisheries, seabirds, and marine mammals. In D. N. Nettleship, G. A. Sanger, and P. F. Springer, eds., *Marine Birds: Their Feeding Ecology and Commercial Fisheries Relationships*. Dartmouth, Nova Scotia: Canadian Wildlife Service, Special Publication.

Furness, R. W. and P. Monaghan. 1986. *Sea Bird Ecology*. London: Blackwell Scientific Publications.

Gentry, R. L. and G. L. Kooyman. 1986. *Fur Seals: Maternal Strategies on Land and at Sea*. Princeton, N.J.: Princeton University Press.

Gotemark, F., D. W. Winkler, and M. Andersson. 1986. Flock-feeding on fish schools increases individual success in gulls. *Nature* 319:589–591.

Graham, F. 1975. *Gulls: A Social History*. New York: Random House.

Grover, J. J. and B. L. Olla. 1983. The role of the rhinoceros auklet in mixed-species feeding assemblages of seabirds in the Strait of Juan de Fuca, Washington. *Auk* 100:979–982.

Harrison. P. 1983. *Seabirds: An Identification Guide*. Boston: Houghton-Mifflin.

Heinrich, B. 1986. Ravens on my mind. *Audobon* 88:74–77.

Hoffman, W., D. Heinemann, and J. A. Weins. 1981. The ecology of seabird feeding flocks in Alaska. *Auk* 98:437–456.

Hunt, G. L., Jr. 1972. Influence of food distribution and human disturbance on the reproductive success of herring gulls. *Ecology* 53:1051–1061.

Hunt, G. L., Jr. and M. W. Hunt. 1975. Reproductive ecology of the western gull: the importance of nest spacing. *Auk* 92:270–279.

Jones, R. E. 1981. Food habits of smaller marine mammals from northern California. *Proc. Calif. Acad. Sci.* 42:409–433.

Kajimura, H. 1985. Opportunistic feeding by the northern fur seal. In J. R. Beddington, R. J. H. Beverton, and D. M. Lavigne, eds., *Marine Mammals and Fisheries*, pp. 300–318. London: Allen and Unwin.

King, J. E. 1964. *Seals of the world.* London: British Museum.

King, J. E. 1983. *Seals of the world,* 2d. ed. Ithaca, N.Y.: Cornell University Press.

Lipps, J. H. and E. Mitchell. 1976. Trophic model for the adaptive radiations and extinctions of pelagic marine mammals. *Paleobiology* 2:147–155.

MacCall, A. D. 1984. Seabird-fishery interactions in eastern Pacific boundary currents. In D. N. Nettleship, G. A. Sauger, and P. F. Springer, eds., *Marine Birds: Their Feeding Ecology and Commercial Fisheries Relationships.* Dartmouth, Nova Scotia: Canadian Wildlife Service Special Publication.

Mate, B. R. 1975. Annual migrations of sea lions *Eumetopias jubatus* and *Zalophus californianus* along the Oregon coast. *Rapp. P.-v. Reun. Cons. int. Explor. Mer.* 169:455–461.

Murphy, E. C., R. H. Day, K. L. Oakley, and A. A. Hoover. 1984. Dietary changes and poor reproductive performance in glaucous-winged gulls. *Auk* 101:532–541.

Peterson, R. S. and G. A. Bartholomew. 1967. *The Natural History and Behavior of the California Sea Lion.* American Society Mammalogists Special Publication No. 1.

Pierotti, R. 1980. Spite and altruism in gulls. *Amer. Nat.* 115:290–300.

Pierotti, R. 1981. Male and female parental roles in the western gull under different environmental conditions. *Auk* 98:532–549.

Pierotti, R. 1982. Habitat selection and its consequences to fitness in the herring gull. *Ecology* 63:854–868.

Pierotti, R., D. G. Ainley, T. J. Lewis, and M. C. Coulter. 1977. Birth of a California sea lion on Southwest Farallon Island. *Calif. Fish and Game* 63:64–66.

Pierotti, R. and C. A. Annett. 1987. Reproductive consequences of dietary specialization and switching in an ecological generalist. In A. C. Kamil, J. R. Krebs, and H. R. Pulliam, eds., *Foraging Behavior,* pp. 217–242. New York: Plenum.

Pierotti, R. and C. A. Bellrose. 1986. Proximate and ultimate causation of egg size and the "thick-chick disadvantage" in the western gull. *Auk* 103:401–407.

Pierotti, R. and D. Pierotti. 1980. Effects of cold climate on the evolution of pinniped breeding systems. *Evolution* 34:494–507.

Pierson, M. O. 1978. A study of the population dynamics and breeding behavior of the Guadalupe fur seal. Ph.D. dissertation, University of California, Santa Cruz.

Quinn, W. H., D. O. Zopf, K. S. Short, and R. T. W. Kuo Yang. 1978. Historical trends and statistics of the southern oscillation, El Niño and Indonesian droughts. *Fish. Bull.* 76:663–678.

Rand, R. W. 1959. The cape fur seal *(Arctocephalus pusillus).* Distribution, abundance, and feeding habits off the southwestern coast of the Cape Province. *S. Afr. Div. Fish. Invest. Rep.* 89:1–28.

Repenning, C. A., R. S. Peterson, and C. L. Hubbs. 1971. Contributions to the systematics of the southern fur seals, with particular reference to the Juan Fernandez and Guadalupe species. In W. H. Burt, ed., *Antarctic pinnipedia (Antarctic Res. Ser.* 18), pp. 1–34.

Repenning, C. A., C. A. Ray, and D. Grigorescu. 1979. Pinniped biogeography. In J. Gray and A. J. Boucot, eds., *Historical biogeography, Plate Tectonics, and the Changing Enviornment,* pp. 357–369. Corvallis. Oregon State University Press.

Richards, F. A. 1981. *Coastal upwelling.* Washington, D.C.: American Geophysical Union.

Ryder, R. A. 1957. Avian-pinniped feeding associations. *Condor* 59:68–69.

Schreiber, R. W. 1970. Breeding biology of western gulls *(Larus occidentalis)* on San Nicolas Island, California, 1968. *Condor* 72:133–140.

Sealy, S. G. 1973. Interspecific feeding assemblages of marine birds off British Columbia. *Auk* 90:796–802.

Sowls, A. L., A. R. DeGange, J. W. Nelson, and G. S. Lester. 1980. *Catalog of California Seabird Colonies.* Washington, D.C.: U.S. Department of the Interior.

Spaans, A. L. 1971. The feeding ecology of the herring gull in the northern part of the Netherlands. *Ardea* 59:75–186.

Tinbergen, N. 1960. *The Herring Gull's World.* New York: Harper and Row.

Vaz-Ferreira, R. 1975a. Behavior of the southern sea lion, *Otaria flavescens*, in the Uruguayan Islands. *Rapp. p.-v. Reun. Cons. int. Explor. Mer.* 169:219–227.

Vaz-Ferreira, R. 1975b. Factors affecting numbers of sea lions and fur seals on the Uruguayan islands. *Rapp. P.-v. Reun. Cons. int. Explor. Mer.* 169:257–262.

Vermeeer, K. 1963. The breeding ecology of the glaucous-winged gull *(Larus glaucescens)* on Mandarte Island. *B.C. Occ. Papers of the British Columbia Prov. Museum* 13:1–104.

Vermeer, K. 1982. Comparison of the diet of the glaucous-winged gull on the east and west coasts of Vancouver Island. *Murrelet* 63:80–85.

Ward, J. G. 1973. Reproductive success, food supply, and the evolution of clutch size in the glaucous-winged gull. Ph.D. dissertation, University of British Columbia, Vancouver.

Wooster, W. S. and J. L. Reid, Jr. 1963. Eastern boundary currents. In M. N. Hill, ed., *The Sea*, pp. 253–280. New York: Wiley Interscience.

Wursig, B. and M. Wursig. 1980. Behavior and ecology of the dusky dolphin, *Lagenorhynchus obscurus*, in the south Atlantic. *Fish. Bull.* 77:871–890.

7

■ Interactions Between Scavenging Seabirds and Commercial Fisheries Around the British Isles

**Robert W. Furness, Anne V. Hudson,
and Kenneth Ensor** · *Department of Zoology
University of Glasgow, Scotland*

Human exploitation of living marine resources has provided an increasing opportunity for some seabirds to take advantage of foods that would otherwise be unavailable to them. Around the British Isles, for example, adult demersal fish such as cod *(Gadus morhua)*, haddock *(Melanogrammus aeglefinus)*, and whiting *(Merlangius merlangus)* do not normally occur in the diet of the seabirds that are capable of diving to the seabed in shallower areas of the sea. Nevertheless, at some localities and at certain times of year, these fish may represent the bulk of the diet, by mass or energy content, of certain seabirds. These are species incapable of diving to the seabed, but able to exploit the activities of commercial fishing boats; species such as great black-backed gulls *(Larus marinus)*, herring gulls *(L. argentatus)*, lesser black-backed gulls *(L. fuscus)*, great skuas *(Catharacta skua)*, and northern gannets *(Sula bassana)*. Kittiwakes *(Rissa tridactyla)* may also make extensive use of this feeding opportunity, and smaller numbers of several other species also join flocks at fishing boats. The only source of demersal fish for these species must be from man's activities, the fish being scavenged from behind commercial fishing vessels or, in the case of some gulls, being stolen after the catches have been landed. In addition, offal (fish livers and intestines) from gutting operations carried out on board may be made available to seabirds around vessels and is particularly sought after by northern fulmars *(Fulmarus glacialis)*.

In many respects the interactions between seabirds foraging at fishing vessels are similar to those between seabirds foraging in flocks over shoals of small fish prey available at the surface. However, studying seabirds at fishing boats has a number of advantages. It is possible to conduct experiments where the availability of food is controlled, and so to examine foraging behavior in relation to identified prey species, quantified prey sizes, and controlled rates of presentation of food. This gives great scope for the investigation of foraging efficiency and the effects of intra- and interspecific competition. It also provides a useful model for the natural relationships between foraging seabirds and fish schools at sea such as discussed elsewhere in this volume.

There has been much dispute over the role of human exploitation of fish and marine mammal populations in causing the observed population changes in seabirds. James Fisher (1952, 1966) was convinced that food provided by man during whaling and trawling operations caused the southward spread and population increase of the northern fulmar in the northern Atlantic. However, his view has been challenged. Wynne-Edwards (1962) suggested that the spread was a natural result of genetic or behavioral evolution, and nothing to do with man. Salomonsen (1965) suggested that it was due to a gradual warming of the northeastern Atlantic which favored the large-billed boreal population of the species. Brown (1970) also felt that oceanographic factors were of prime importance since he found little association between northern fulmar distribution in the northwest Atlantic and the distribution of the fishing fleet. Dietary studies at Scottish colonies of northern fulmars showed an enormous difference in food types taken by breeding birds at a colony to the west of Scotland compared with food types taken by breeding birds at a colony to the north of Scotland. Many of the food items important in the diets at these locations were not obtained from fishing boat activities (Furness and Todd 1984). Studies of the densities of northern fulmars through the North Sea (Blake et al. 1984) provided only weak evidence for an influence of fishing boat distribution on northern fulmar distribution at sea.

Rather little attention has been paid to the ways in which seabirds exploit fishery waste, and competitive interactions between different scavenging seabird species and age classes, the amounts of offal and discarded whole fish made available to seabirds, or the proportion of this that is exploited. Recent work in Queensland, Australia from prawn trawlers in Moreton Bay indicates that scavenging seabirds (gulls, terns, and cormorants by day and gulls and terns at night) obtain about 30–35 percent of discarded fish during the day and about 16 percent at night, while dolphins account for about 40–50 percent of the discards, and the remainder sink (T. J. Wassenberg, personal communication). Rees (1963) observed scavenging behavior of seabirds in the Gulf of St. Lawrence and the Strait of Belle Isle, and Rodriguez (1972) observed scavenging seabirds at fishing boats off the southwestern coast of Africa. According to the observations of Rodriguez (1972), some 125 freezer trawlers fished in June–July 1967, predominantly for hake *(Merluccius)*, in the 444-by-222 kilometer rectangle bounded by 23°-27°S, 13°–15°E, and discarded a total of 576 metric tons of waste fish and offal daily. Much of this was consumed by huge numbers of associated scavenging seabirds, of which an estimated total of 46,000 albatross and large petrels predominated, particularly black-browed albatross *(Diomedea melanophris)* and white-chinned petrels *(Procellaria aequinoctialis)*, which fed almost exclusively on the waste. Rodriguez describes the clearly defined differences in the positions behind trawlers taken by each species apparently as a result of differences in competitive abilities and flight characteristics. Abrams (1983) also concluded that the availability of trawler waste had influenced the distributions of scavenging seabirds, particularly black-browed albatross, in the Benguela Current off South Africa and concluded that scavenging behind trawlers had been a factor causing an increase in numbers of black-browed albatross as well.

In this paper we will consider two topics. First, how much food is made available to scavenging seabirds by the fishing industry around the British Isles and to what extent is this utilized by seabirds? Second, how do the scavenging seabirds interact when feeding at fishing boats and do the interactions suggest that the population dynamics of scavenging seabirds may be affected by competition for food at fishing boats?

■

Methods

Study Area and Fish Catch Data

The study area is defined by the division of the North Atlantic and North Sea as indicated in figure 7.1. Based on the Food and

Agriculture Organization (United Nations) map, and used to delineate the International Council for the Exploration of the Seas (ICES) fishing areas. Quantities of each species of fish and shellfish harvested in each of these areas are listed in the statistical bulletins of the ICES, which are published annually. The most recent bulletin, published in 1985 and dealing with fish catches in 1982, has been used as the primary source of information from which we have derived estimates of the amount of food (discards and offal) made available to seabirds in each of the areas in figure 7.1. Comparison

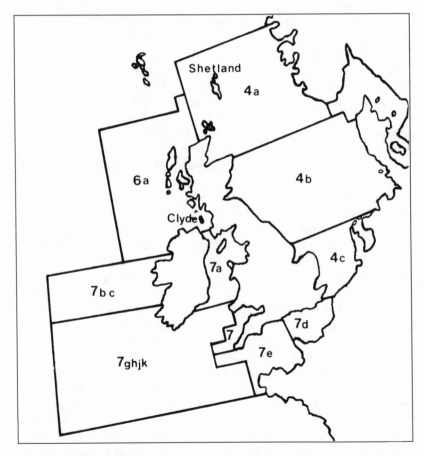

Figure 7.1. ICES fishing areas around the British Isles: 4a-northern North Sea; 4b-central North Sea; 4c-southern North Sea; 6a-northwest Scotland; 7a-Irish Sea; 7bc-western Ireland; 7de-English Channel; 7f-Bristol Channel; 7ghjk-Southwest Approaches.

with volumes for the years 1977 to 1981 shows that quantities have been fairly consistent between years so that the 1982 data are taken to be indicative of the general pattern for the years immediately before, and probably after, 1982.

Discard and Offal Quantities

In order to estimate consumption of fishery waste by seabirds we need to know how we can use data on fish landings in order to assess the quantities of offal and discards (fish of no commercial value plus fish of commercial value that are of a size not worth landing or below the legal size limit) that are thrown back overboard by fishermen and so made available to scavenging seabirds. Quantities of offal and discards produced depend firstly on the fishing method employed and the ways in which the fish are processed, so that observations were made on a number of different types of fishing vessels. Clearly it is not possible to make observations, as ideally required, from all types of fishing vessels of each size and of each nationality in all areas around the British Isles in each session of the year over several years. We have been limited to sampling from British vessels, principally of short trip duration (generally one-day fishing trips from Shetland, the Clyde Sea area, and northwest Scotland), and we have assumed that discarding practices are comparable among vessels of other countries and in other areas around Britain and Ireland. This assumption may not be correct, and so our estimates of quantities of offal and discards may need to be revised if foreign vessels treat their catches appreciably differently from the way that they are processed by British boats.

We also need to determine how much of the available offal and discards is consumed by seabirds. These questions were addressed by making observations from trawlers around the Shetland Islands and in the Clyde Sea area and around the Small Isles (Rhum, Eigg, Muck, and Canna) in northwest Scotland (figure 7.1) during their normal fishing activities. We gutted a number of fish of various species in order to estimate the weight of offal as a proportion of fish weight, allowing the total quantity of offal to be calculated from total landing figures. The sizes of catches made at each haul, and quantities of fish discarded, were estimated in terms of boxes of fish, each boxful weighing on average 45 kilograms (Hudson 1986). The quantity landed at market was recorded and this, once corrected to allow for the weight of offal removed, provided an independent check of our estimate of the number of boxes of marketable fish, since this gave the quantity of fish discarded at sea by subtraction from the estimated total catch.

Seabirds Using Fishery Waste

During each fishing trip we counted numbers of seabirds associating with the vessel and behind other vessels; we also observed boats from the mainland and, in the vicinity of Shetland, from a light aircraft. Although results differ between boats of different sizes and fishing methods, between areas and at different times of year, constraints on fieldwork meant that we had to concentrate effort at sea largely on small whitefish trawlers around Shetland in summer and on Norway lobster *(Nephrops norvegicus)* boats in the Clyde Sea area and around the Small Isles of northwest Scotland throughout the year (figure 7.2). Other studies have dealt with other types of boat and other areas (Irish Sea: Hillis 1971, 1973 and Watson 1981; North Sea: Anstey 1984 and Blake et al. 1984; northwest

Figure 7.2. Gulls dropping onto discarded fish behind a trawler in the Clyde Sea Area.

Scotland: Boswall 1960, 1977 and Lockley and Marchant 1951; southwest England: Lake 1984).

In order to convert quantities of discards and offal available to seabirds into a measure of the numbers of birds that may be sustained by scavenging at boats we measured the energy value of offal and discards by bomb calorimetry and estimated the energy requirements of seabirds from bioenergetics equations following the results presented by Ellis (1984).

Feeding Behavior and Interactions

The ability of different age classes and species of seabirds to exploit offal and discards was assessed by experimental discarding performed by ourselves while fishermen were engaged in normal fishing activities (i.e., sorting, gutting, and/or discarding, and/or trawling for the next catch). Individual fish that were about to be discarded by the fishermen were taken, identified as to species, measured, and then discarded. Thus our observations of seabirds taking measured fish were made while birds were also feeding on fish being discarded in the normal way by fishermen. Attempts by seabirds to obtain each individual fish were filmed using a JVC KY1900EL10 video camera with a synchronized sound track and a JVC CR4400E portable recorder. Subsequently, when the methodology had become routine, the video equipment was replaced by a portable tape recorder and stopwatch system which allowed data to be transcribed much more quickly with little loss of accuracy or information. The following data were recorded: fish species; fish length (mm); species and age (where this could be determined from plumage) of the seabird attempting to take the fish; the outcome of the attempt (fish missed, swallowed, dropped, or stolen); the time taken between fish being grabbed by the bird and either swallowed, dropped, or stolen; whether the bird swallowed the fish on the water or in the air; if birds fought for the fish; the species and age of kleptoparasite; and whether the fish was eventually swallowed or sank. Because some fish were tackled by a series of birds the sequence could occasionally be complex and could only be recorded by video film or by describing events and recorded stopwatch timings into a tape recorder.

Swennen and Duiven (1977) found that the maximum body width and depth (or cross-sectional area) of fish was the key measurement used by alcids in choosing optimal prey sizes. This may be a more important feature of a fish than its length for prey selection by scavenging seabirds as well, but it is difficult to measure maximum fish depth or breadth as accurately or as easily as one can measure fish length, and both depth and breadth measures tend to be influenced by the handling of the fish during capture, since these measurements are more sensitive to effects of pressure change and compaction which result, for example, in the stomachs of many of the fish being forced out through the mouth or the swim bladder ruptured. Also, relationships between fish length and weight or otolith length have been published for many marine fish, but few exist allowing fish breadth or depth to be estimated from otolith dimensions. In this paper we have considered only fish length, as measured, since within a given fish species, fish breadth or depth is closely related with the length of the fish.

Behavioral records were analyzed by use of the SPSS and SPSS-X packages on the Glasgow University ICL 2988 computer. Where statistical tests have been performed, significant means $p < 0.05$.

■

Results

Fish Discards and Offal Made Available

Observations from fishing boats catching sand eels *(Ammodytes marinus)* or other fish for industrial purposes (reduction to oil and fish meal) showed that these fisheries provide very little food for seabirds. Gulls do attend sand eel boats and pick fish from the net when it is hauled, but the total quantity of food obtained is very small by comparison with amounts available from whitefish boats. Similarly, boats fishing for herring *(Clupea harengus)* or mackerel *(Scomber scombrus)* do not normally provide seabirds with feeding opportunities, although on occasion quotas can result in discarding of these species. Scavenging seabirds obtain most fishery waste from whitefish and Norway lobster boats.

All areas around the British Isles except for the Irish Sea are dominated by catches of industrial species, or of herring or mackerel, so that total fish catch data provide little indication of the availability of fishery waste to seabirds.

Observations from whitefish and Norway lobster boats indicate that the amounts of fish discarded vary greatly between catches, between boats, between areas, and seasonally. Our data and data of

248 Robert W. Furness, Anne V. Hudson, and Kenneth Ensor

the Department of Agriculture and Fisheries for Scotland (DAFS) suggest that variation between catches is almost as great as variation between areas and seasons (table 7.1), so we have used the same factors to estimate discard quantities from landings data for each fishery region. We have assumed that, for whitefish boats, the discard mass averages 20 percent of the mass of landed haddock and whiting, 10 percent of the mass of landed hake *(Merluccius merluccius)*, ling *(Molva molva)*, pollack *(Pollachius pollachius)*, tusk *(Brosme brosme)*, monkfish *(Lophius piscatorius)*, gurnards (Triglidae), and 5 percent of the mass of landed cod and saithe *(P. virens)*. Norway lobster boats, on average, discard about twice as much fish as the quantity of *Nephrops* caught (table 7.1). These values may be exceeded in certain years when good cohorts recruit into the fishery (for example, DAFS data indicate that as much as 42 and 48 percent of haddock weight caught by seine net vessels in 1975 and 1980 was

Table 7.1. Quantities (weights) of fish discarded from fishing boats.

Fishing Method	Locality	Number of Catches Sampled	Average Rate of Discarding Fish	Source
Nephrops trawl	Clyde	18	1.8 × mass of *Nephrops* catch (range 0.5 × to 3.0 ×)	This study
	Hebrides	9	2.2 × mass of *Nephrops* catch (range 1.0 × to 3.0 ×)	This study
Seine net	North Sea	> 50	33% of haddock caught 25% of whiting caught 10% of cod caught	Jermyn and Robb (1981)
Motor trawl	North Sea	> 20	20% of haddock caught 20% of whiting caught 1% of cod caught	
Light trawl	Shetland	151	27% of whitefish caught (range 1% to 85% of catch)	Hudson (1986)

NOTE: Discard rates are given either as mass of fish as a multiple of the measured quantity of *Nephrops* landed by *Nephrops* trawlers or as a percentage of the total mass of whitefish or whitefish species caught by whitefish boats. Sample size is the number of separate catches that were examined.

discarded because of the strong 1974 and 1979 cohorts. Thus, in some years the total amount of discards may be elevated by such occurrences. However, good years for one fish species may not be good for others, so that total discard amounts will fluctuate much less than the discard amounts of particular species.

By applying these values to ICES landings data for 1982 we can obtain a crude estimate of the annual total mass of discards in each fishery area (table 7.2). From analyses of the mass of offal in relation to fish mass (table 7.3) we can convert the same fish landings data to quantities of offal discharged, on the assumption that all of these species are gutted at sea and that all offal is discharged (table 7.2). In practice, some fish are landed whole and a few boats may retain offal, but these practices probably have little effect on the overall calculations.

In the Irish Sea, Norway lobster boats are the main source of discard fish, while total discard quantities are much the highest in the north and central North Sea. Offal mass is also much greater in these last two regions than elsewhere (table 7.2).

Norway lobster trawlers use a much smaller mesh net than whitefish trawlers and as a result the fish discarded from Norway lobster trawlers tend to be smaller than those discarded from whitefish trawlers. Around Shetland, most discards were of had-

---

Table 7.2. Quantities of offal and fish discards from whitefish and Norway lobster boats around the British Isles.

| | | Whitefish Discarded | |
ICES Fishing Ground	Offal Discarded (metric tons)	Whitefish Boats (metric tons)	Nephrops Boats (metric tons)
Northern North Sea	40,000	41,000	3,000
Central North Sea	34,000	30,000	8,800
Southern North Sea	9,700	5,300	10
Northwest Scotland	10,000	13,000	18,000
Irish Sea	3,700	1,400	20,000
Western Ireland	1,600	1,800	2,500
Eastern English Channel	1,200	1,000	0
Western English Channel	1,900	900	20
Bristol Channel	1,000	1,100	10
Southwest Approaches	4,300	4,500	7,300

NOTES: Data for 1982; for methods of calculation and assumptions made see text.

dock and whiting with median lengths of 28 and 29 centimeters respectively (Hudson 1986). Discards from Norway lobster boats in the Clyde and around the Inner Hebrides tended to be predominantly 10–20-centimeter flatfish, 10–25-centimeter whiting, Norway pout *(Trisopterus esmarkii)* or haddock (Ensor and Furness, unpublished data). In the Irish Sea most discards are whiting, dab *(Limanda limanda)*, poor cod *(Trisopterus minutus)*, or Norway pout, and are generally 5–24 centimeters in length (Watson 1981).

Calculations of the Numbers of Seabirds Supported by Fishery Waste

The best measurements of the daily energy expenditure of seabirds are recent studies using isotopically labeled water. Ellis (1984) reviewed seven studies of seabirds where such methods indicated daily expenditures of 2.6, 3.0, 3.1, 3.3, 3.4, 4.8, and 5.2 times Basal Metabolic Rate (BMR). Subsequent studies (Gaston 1985; Roby and Ricklefs in press; Birt et al. unpublished data) also indicate that the daily energy expenditure of seabirds is generally between 3 and 4 times BMR. Since most of these studies are of breeding seabirds, and nonbreeders have a lower energy requirement (estimated to be 20 to 30 percent lower in thick-billed murres *(Uria lomvia)* and black guillemots *(Cepphus grylle)*; Gaston 1985) we have chosen 3 BMR as a suitable estimate of the daily energy requirement of seabirds over the whole year. According to Ellis (1984) seabirds from

Table 7.3. Offal mass as a proportion of total fish mass.

Source of Data	Relationship Given
This study	Offal = 11.0% of mass of gadoid fish caught
This study	Offal = 6.5% of mass of flatfish caught
DAFS[a] (A. S. Jermyn, personal communication)	Offal = 11.1% of mass of fish caught Offal = 12.5% of mass of gutted fish
Boswall (1960)	Offal = 12.5% of mass of fish caught Offal = 14.3% of mass of gutted fish
Bailey and Hislop (1978)	Offal = 10–15% of gadoid body mass

[a] Department for Agriculture and Fisheries (Scotland).

high latitudes have a greater BMR than tropical seabirds. Herring gulls and great skuas have a measured BMR 1.25 and 1.27 times that predicted by an allometric equation based on data for all seabirds. We assume a BMR in line with these deviations from a common regression, which gives a value of 478 kilojoules per day for a 1,000-gram seabird.

From measurements of the calorific value of fish offal and tissues (table 7.4), we have calculated on the basis of offal containing 11 kilojoules g^{-1} and discards 5 kj g^{-1}. Assuming a food utilization efficiency of 75 percent (derived from Kendeigh, Dolnik, and Gavrilov 1977) we can estimate that a 1,000-gram seabird could survive on 64 kilograms of offal per year or on 120 kilograms of discards per year. Most scavenging seabirds around the British Isles are northern fulmars or herring gulls which weigh close to 1,000 grams, while great skuas and great black-backed gulls weigh from 1,200 to 1,800 grams. Kittiwakes at 360 grams and northern gannets at 3,000 grams are rather far from a 1,000-gram standard, but represent relatively smaller numbers of the birds obtaining food at fishing boats.

By combining these data with the quantities of offal and discards available (table 7.2), we can estimate the maximum number of 1,000-gram seabirds that could be sustained by offal and discards in each area (table 7.5). In theory, if all of the offal and discards were taken, then some 2.8 million 1,000-gram seabirds might be supported by this food supply. However, this total exceeds the number

Table 7.4. Caloric value of fish offal and discards.

Sample Analyzed	Fish Length (cm)	Sample Size	Caloric Value (kJ g^{-1})	Source
Offal (saithe)	28–29	3	7.3	this study
Offal (whiting)	18–28	6	12.6	this study
Saithe (whole)	28–30	2	5.2	this study
Whiting (whole)	18–28	5	5.8	this study
Saithe (whole)	3–4	8	5.1	Harris and Hislop (1978)
Whiting (whole)	4–6	3	4.1	Harris and Hislop (1978)
Saithe (gutted)	29	1	4.9	this study
Whiting (gutted)	18–28	5	4.4	this study
Cod (fillet)	—	3	3.2	Paul and Southgate (1975)
Haddock (fillet)	—	—	3.1	Paul and Southgate (1975)
Plaice (whole)	—	8	3.8	Paul and Southgate (1975)

of seabirds that could actually be supported. In practice, most offal is taken, although on the occasions when large quantities of offal are washed out of the scuppers at once, much tends to sink before seabirds can reach it. Some discards are too large to be handled or are of species that are not easily swallowed (see below), so that the available offal and discards could only support somewhat smaller numbers than these calculations suggest.

Seabirds Using Waste in Different Areas

The numbers of each species of seabird in flocks foraging at fishing boats vary considerably between areas and to some extent between seasons (table 7.6).

Foraging Efficiency of Different Age Classes

In the Clyde Sea area juvenile herring gulls missed (i.e., failed to get hold of) 4.4 percent of the discards they attempted to pick up. Significantly fewer were missed by older birds: one-year-old birds

Table 7.5. Numbers of 1,000-gram seabirds that could be supported by offal and discards around the British Isles.

	1,000-gram Seabirds Potentially Supported By	
Fishing Ground	Offal	Discards
Northern North Sea	620,000	310,000
Central North Sea	530,000	277,000
Southern North Sea	150,000	38,000
Northwest Scotland	172,000	220,000
Irish Sea	58,000	153,000
Western Ireland	25,000	31,000
Eastern English Channel	19,000	7,000
Western English Channel	30,000	6,000
Bristol Channel	15,000	8,000
Southwest Approaches	67,000	84,000
All areas	1,686,000	1,134,000

Table 7.6. Numbers of seabirds associated with fishing boats in different areas.

Seabird	Irish Sea		Clyde Sea	Shetland Summer		Eastern Scotland Winter	Northwest Scotland Summer
	Summer (18)	Winter (32)	Winter (149)	Inshore (72)	Offshore (40)	( )	(1)
Northern fulmar	23 (0–100)	3 (0–30)	0 (0–3)	485 (10–2,500)	721 (0–6,000)	1 (0–4)	155
Northern gannet	11 (0–50)	6 (0–50)	6 (0–167)	9 (0–200)	3 (0–100)	15 (0–51)	5
Herring gull	213 (0–800)	251 (10–900)	241 (0–1,340)	30 (0–400)	0	87 (0–380)	15
Lesser black-backed gull	2 (0–10)	0 (0–1)	3 (0–88)	6 (0–20)	0	2 (0–28)	1
Great black-backed gull	5 (0–30)	16 (0–60)	6 (0–80)	234 (10–1,000)	19 (0–300)	2 (0–24)	90
Kittiwake	75 (0–200)	78 (2–400)	24 (0–285)	3 (0–50)	0	0	35
Great skua	0 (0–6)	0 (0–6)	0 (0–1)	12 (0–50)	1 (0–20)	0	0

SOURCES: Irish Sea: Watson 1981; Clyde Sea and Shetland: this study; eastern Scotland: Anstey 1984, northwestern Scotland: Boswall 1960.
NOTES: Summer = April–July, winter = October–February; numbers of counts are given in parentheses below the location heading; minimum and maximum counts are given in parentheses below means except where all counts were zeros.

missed 0.8 percent while two-year-olds and older birds missed only 0.4 percent (n = 1,365 fish, x^2_2 = 26.7, p < 0.005). In addition, juvenile herring gulls were more likely than adults to drop the fish they did pick up (9.7 percent dropped by first-year birds versus 6.2 percent by adults, x^2_2 = 3.97, p < 0.05).

Adult herring gulls in the Clyde obtained more fish per bird than did immature herring gulls (table 7.7). Adult herring gulls were also more selective in the sizes of fish they took. While mean fish lengths were almost the same (first year birds, 24.6 cm; two- to-four-year-olds, 24.5 cm; adults, 24.9 cm), the variance of fish length was greater for first-year birds (13.76) than for two-to four-year-old birds (11.70) or adults (8.46). Adults were less likely than juveniles or immature birds to take particularly small or particularly large fish = 1.626, p < 0.001, $F_{448,404}$ = 1.383, p < 0.05).

Around Shetland in summer the largest numbers of discard fish were taken by great black-backed gulls, many of which were immature birds. For both adults and immature birds the handing time of fish (time taken from grabbing the fish to swallowing it) increased at an accelerating rate with fish length. Handling times of immature birds were longer for all fish sizes than those of adults and the relationships between log handling time and log fish length were linear (fig. 7.3).

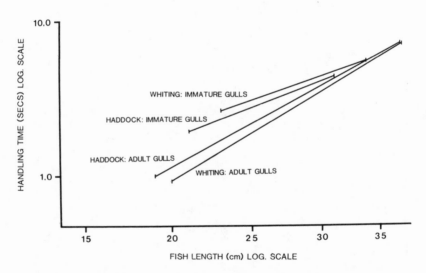

Figure 7.3. Log. handling times (secs.) of haddock and whiting by adult and immature great black-backed gulls at trawlers around Shetland in relation to log. fish length (cm).

Table 7.7. Numbers of discard fish swallowed by scavenging seabirds behind *Nephrops* trawlers in the Clyde in relation to numbers predicted from the relative numerical abundance of each species and age class.

Species	Age	Mean Number Present per Boat	Fish Swallowed	Expected Number Swallowed[a]	Relative Foraging Success
Northern gannet	Adult	5.7	53	25.9	2.05
Great black-backed bull	All	6.0	35	27.3	1.28
Herring gull	Adult	78.5	405	357.9	1.13
Herring gull	Immature	162.7	742	742.1	1.00
Lesser black-backed gull	All	2.6	11	11.9	0.92
Kittiwake	Adult	21.8	27	99.4	0.27
Kittiwake	Immature	2.3	1	10.6	0.09

NOTES: Null hypothesis that birds of all species and age have equal likelihood of obtaining each fish; relative foraging success is calculated as number of fish swallowed divided by expected number.
[a] Derived from the relative abundance of the birds.

Foraging Efficiency of Different Species

When offal was discarded from Shetland trawlers virtually all of it was consumed by northern fulmars. The large numbers and aggressive behavior of northern fulmars generally precluded other species from entering the area close beside the boat where offal was discharged (figure 7.4). In areas where northern fulmars were rarely found (e.g., the Clyde Sea) offal was taken by gulls in preference to discard fish, as might be expected since it has a considerably higher calorific value. Northern fulmars have great difficulty swallowing whole fish and around Shetland they tended to consume offal but ignore fish. Although northern fulmars outnumbered gulls, skuas, and northern gannets around inshore Shetland trawlers by a factor of 1.65 to 1 (table 7.6), they swallowed only 88 of the fish experimentally discarded, compared to the 3,705 swallowed by gulls, skuas, and northern gannets. Thus northern fulmars represented 62.3 percent of all the birds present but obtained only 2.3 percent of the discards swallowed.

Comparing the foraging success of adult seabirds taking discards behind *Nephrops* trawlers in the Clyde Sea area, northern gannets swallowed all 53 fish they attempted to take (100 percent; lesser

Figure 7.4. Fulmars scavenging offal around a trawler in Shetland.

black-backed gulls obtained 9 in 9 attempts (100 percent); great black-backed gulls obtained 21 in 22 attempts (95 percent); herring gulls obtained 405 in 477 attempts (85 percent); and kittiwakes obtained 27 in 32 attempts (84 percent). Apart from 3 fish missed by herring gull adults, the failure to swallow fish was due either to the fish being dropped or to it being stolen by a kleptoparasite.

Although kittiwakes obtained 84 percent of fish that they attempted to take, the species took a much lower proportion of all fish discarded than would have been predicted from the numbers of each seabird species present (table 7.7). Numerically, immature herring gulls predominate around boats in the Clyde and took fish in the proportion predicted by their abundance relative to all seabirds present. Northern gannets took twice as many fish as their numbers would have predicted. Great black-backed gulls and adult herring gulls also obtained more than their numerical abundance would have predicted. The differences between observed and expected numbers taken by each species and age class are highly significant $(X^2_6 = 98.1, p < 0.001)$.

One reason for kittiwakes and lesser black-backed gulls only attempting to take a fraction of the number of fish taken by larger species is that they are unable to handle and swallow large fish. The mean length of fish swallowed by each species is less than the mean length of fish dropped, and increases with seabird size (table 7.8).

The handling times of all seabirds around Shetland increased at an accelerating rate as fish length increased (figure 7.5). Northern gannets and great black-backed gulls were considerably quicker at swallowing fish than herring gulls or great skuas and also managed to swallow larger fish. Northern fulmars were particularly slow to

Table 7.8. Mean lengths of fish taken by seabirds around *Nephrops* trawlers in the Clyde.

Species	Fish Swallowed			Fish Dropped		
	N	mean (cm)	s.d.	N	mean (cm)	s.d.
Northern gannet	53	28.3	3.6	0	—	—
Great black-backed gull	35	29.0	3.4	2	32.5	0.7
Herring gull	1,147	24.7	3.2	100	27.6	2.7
Lesser black-backed gull	11	21.9	4.9	1	26	—
Kittiwake	27	15.8	2.9	5	16.4	4.6

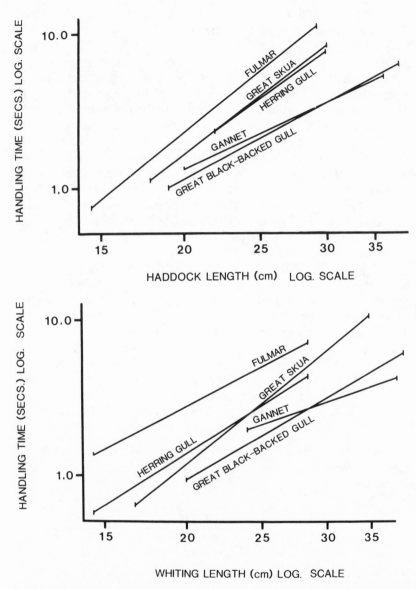

Figure 7.5. Log. handling times (secs.) of haddock and whiting by adult northern gannets, great black-backed gulls, great skuas, herring gulls, lesser black-backed gulls, and northern fulmars at trawlers around Shetland in relation to log. fish length (cm).

swallow fish. These differences in handling times between species were statistically significant.

Feeding Interactions

Although great skuas have the reputation of being specialist klep-toparasites, most stealing of discards in foraging flocks at trawlers around Shetland was done by great black-backed gulls, and great skuas lost more fish to kleptoparasitism than they obtained in this way (table 7.9). In general, larger species of seabirds tended to steal more than smaller seabirds and this appears to be related to the difficulty that the smaller species had in swallowing the larger discarded fish. The mean length of fish stolen from great skuas (28.3-cm haddock and 29.9-cm whiting) were significantly larger than the mean lengths of these fish swallowed by great skuas (25.5-cm haddock and 27.4-cm whiting;. t = 7.7 and 7.2 respectively, p < 0.01). Kleptoparasitism appears to provide an important part of the total discard consumption of great black-backed gulls (table 7.9) and is also an important cause of loss of fish for herring gulls and lesser black-backed gulls.

In the Clyde, kittiwakes were never, and northern gannets rarely, seen stealing fish. Herring gulls were less often kleptoparasitic than were lesser black-backed gulls or great black-backed gulls, although because of their numerical predominance in the area most fish were stolen by herring gulls (the proportions of discarded fish swallowed that were obtained by kleptoparasitism increased as follows: kitti-wakes 0.0 percent, northern gannets 1.9 percent, herring gulls 7.9 percent, lesser black-backed gulls 9.1 percent, great black-backed gulls 12.9 percent: the total was 1,277 fish swallowed).

As a consequence of size preferences and kleptoparasitic interactions most discarded fish were consumed by seabirds, although flatfish, gurnards, and roundfish more than 35 centimeters in length tended to be ignored unless nothing else was available. Overall about 75 percent of all experimental discards were taken by seabirds both around Shetland and in the Clyde and around the Small Isles. This proportion probably applies to the fish discarded by fishermen too, except where catches include high proportions of flatfish, which are often ignored by seabirds and also tend to sink faster than roundfish.

■

Discussion

The quantities of offal and discarded fish made available to scavenging seabirds in the sea areas round the British Isles (figure 7.1) are enormous and could possibly support as many as 2.5 million 1,000-gram seabirds (table 7.5; allowing for 25 percent of discards being lost to seabirds). Data in the ICES statistical bulletins indicate that there is little seasonal variation in total catches of fish. This is probably the case also for provision of offal and discards, although there is a brief hiatus over the Christmas and New Year period when most fishermen are at home. We cannot say at present whether seabird requirements for offal and discards is greater in summer or winter. Some species are migratory, while the availability of alternative foods (e.g., surface shoals of sand eels) also varies between seasons.

Clearly seabirds cannot consume all of the offal and discards made available by fishing boats. Some must sink or be taken by fish or marine mammals. However, our observations in the Clyde, Inner

Table 7.9. Kleptoparasitic interactions involving scavenging seabirds at trawlers around Shetland.

Species	Number of Fish Obtained by Kleptoparasitism	Number of Fish Lost to Kleptoparasitism	Fish Obtained or Lost as a Consequence of Kleptoparasitism[a] Obtained	Lost
Great black-backed gull (adult)	335	173	12.7	6.5
Great black-backed gull (immature)	44	36	9.6	7.9
Northern gannet	58	40	10.8	7.5
Great skua	94	135	17.2	24.6
Herring gull	5	58	2.2	25.9
Lesser black-backed gull	0	27	0.0	32.1
Northern fulmar	2	69	1.3	43.4

[a](As a percentage of all discarded fish handled by the species.)

Hebrides, and around Shetland indicate that very little offal is not obtained by seabirds, and few discards other than flatfish or round-fish of more than 35 centimeters in length are not consumed by birds. Probably about 90 percent of offal and 75 percent of discards are taken. Off eastern Canada, fishermen clean whitefish by cutting off the head and throwing it into the sea with the guts attached. This rapidly sinks and so seabirds are able to obtain only a small part of the offal before it has sunk too deep for them to reach it (R. G. B. Brown, personal communication). This simple difference in gutting method (British fishermen remove the guts from the slit-open abdominal cavity without taking the head off the fish) is a major determinant of the ability of scavenging seabirds to exploit offal. Around the British Isles, offal thrown overboard by fishermen tends to float, since the liver in particular is buoyed up by its high lipid content.

Recent estimates of seabird population sizes in northwest Europe (Evans 1984) indicate that there are well over 3 million scavenging seabirds in the fishery areas around the British Isles. Estimates of the breeding populations in the British Isles alone are of 600,000 northern fulmars (breeding individuals), 700,000 herring gulls, 300,000 northern gannets, 50,000 great black-backed gulls, 100,000 lesser black-backed gulls, 1.1 million kittiwakes, and 20,000 great skuas. There is probably an equal number of immature birds of each of these species. Further, large numbers of herring gulls breed on the continental coast of Europe. In addition, large numbers of birds from populations in Norway, Iceland, and the Faeroe Islands spend the winter around the British Isles and some nonbreeders may re-main in British waters all year. A large proportion of the herring gulls around Shetland and on the east coast of Britain in winter are from an estimated 200,000 pairs breeding in northern Norway (Coulson et al. 1984), while the numbers of great black-backed gulls (particularly immature gulls) that we recorded around Shetland in summer greatly exceeded the number that could be attributed to British breeding colonies alone, and presumably originate from the much larger populations of that species found in Norway (an esti-mated 60,000 pairs; Evans 1984). Since the Norwegian local fishery for demersal fish is principally one of longline fishing for cod, which results in little or no discarding, it is perhaps not surprising that immature Norwegian gulls may prefer to feed around Shetland where discard volumes are very large (table 7.2).

Because the estimates of total seabird population sizes (including nonbreeders) are not very precise, and knowledge of their movements is limited, and because the estimates of the total mass of offal and discards are very crude, we cannot infer that there is not enough offal and discard fish to satisfy the requirements of scavenging seabird populations, but the data suggest that seabird numbers considerably exceed the estimated 2.5 million or so that could be supported on fishery waste alone.

Clearly there are differences in both the geographical distributions and habitats used by each scavenging seabird species. Northern fulmars predominate around Shetland and northwest Scotland, while herring gulls predominate in the Irish Sea, the Clyde Sea, and off eastern Scotland. While northern fulmars represent two-thirds of the seabirds behind trawlers inshore around Shetland in summer, they represent over 95 percent of the seabirds behind offshore trawlers. Interestingly, great black-backed gulls show a greater tendency to utilize the offshore habitat than do herring gulls (table 7.6), a pattern also found on the Grand Banks and Scotian Shelf (R. G. B. Brown, personal communication).

Our observations of the feeding behavior of scavenging seabirds at boats indicate that there is indeed considerable competition for food. There is a clear dominance hierarchy, with northern fulmars at the apex, able to obtain the choicest pickings which are clearly fish livers or the entire offal. For anatomical reasons northern fulmars are unable to swallow whole fish efficiently unless these are particularly small, so that northern gannets and adult great black-backed gulls obtain the pick of the discard fish and generally achieve the highest feeding rate on discards (table 7.7). Immature great black-backed gulls are less efficient than adults (figure 7.2), a situation found for herring gulls foraging at garbage dumps (Burger and Gochfeld 1981; Greig, Coulson, and Monaghan 1983).

Great skuas, although often attempting to steal fish by kleptoparasitism, are unable to swallow many of the large fish taken by great black-backed gulls and as a result they are robbed of large fish more often than they obtain fish by robbery (table 7.9). Herring gulls are outcompeted by great skuas, great black-backed gulls, and northern gannets around Shetland. In the Clyde and Irish seas herring gulls predominate numerically and do not have to compete with large numbers of the larger seabirds. Fish discarded from Norway lobster trawlers are smaller and so can be utilized by herring gulls and to some extent by kittiwakes, which are generally unable to obtain food at boats around Shetland. However, kittiwakes have difficulty competing with herring gulls and their foraging success at boats is low (table 7.7), partly because they drop more fish than do larger birds (table 7.8).

Kleptoparasitism tends to result in unusually large fish being stolen, but it is not clear whether kleptoparasites select victims on the basis of the size of fish they carry (as suggested by Brockman and Barnard 1979), or whether the longer handling times of birds with large fish make them more vulnerable to kleptoparasitism. Although great skuas lost more fish to kleptoparasites than they stole (table 7.9) they were the most kleptoparasitic of the species around boats in Shetland, obtaining 17.2 percent of their fish by kleptoparasitism compared to 12.7 percent stolen by adult great black-backed gulls and 9.6 percent stolen by immature great black-backed gulls. This is consistent with the idea that great skuas are more specialized kleptoparasites than gulls (Furness in press). The lower foraging success of great skuas reflects both the problems they have handling large fish and the fact that they are greatly outnumbered by great black-backed gulls around Shetland boats.

The fact that offal availability is very high in the north and central North Sea and northwest Scotland by comparison with other areas is interesting since it is in these areas that northern fulmar numbers are highest. The low availability of offal further southwest might have been one factor inhibiting northern fulmars from colonizing these regions in large numbers. However, the extent to which discards and offal may influence population dynamics is unclear. Breeding northern fulmars in Shetland fed extensively on sand eels when these were abundant (Furness and Todd 1984). In the last three summers, industrial catches, and apparently recruitment, of sand eels around Shetland have been declining, with the 1985 year class considered to be the poorest since industrial fishing for sand eels began around Shetland in 1974, and so the poorest cohort on record (Gauld, McKay, and Bailey 1986). Sand eels have not been prominent in northern fulmar diets (Hudson 1986), suggesting that the extent to which northern fulmars make use of offal may depend in part on the availability of other foods. Study of great skua diets between 1973 and 1977 indicated that discards were taken as the principal food of nonbreeders, but that breeders took a higher proportion of sand eels and fed the chicks predominantly on sand eels (Furness and Hislop 1981). Since 1981 great skua diets have contained a much lower proportion of sand eels, which also reflects the reduction in sand eel recruitment, and probably in stock size. In addition, the tendency for nonbreeders to make more extensive use of fishery waste appears to be a general one. Obtaining fish from boats probably requires more time but less effort than catching shoaling fish such as sand eels. Since gliding along behind a fishing boat is probably not energetically costly for a gull or a skua (perhaps costing about 3 BMR; Furness and Monaghan 1987), while flying rapidly out to a shoal of sand eels and competing for position in a feeding flock above the shoal will cost as much as or more than the cost of sustained flapping flight (about 10 BMR; Furness and Monaghan 1987), then the energetic cost of following fishing boats to await an opportunity to obtain discards and offal may be no more than existence costs and may result in a better reward to nonbreeders than attempting, at high energy cost, to compete with adults at sand eel shoals. Adults may prefer to expend this extra energy in order to minimize the time spent foraging so that they can feed their chicks rapidly and spend more time guarding them.

Although numbers of all scavenging seabirds in the British Isles have increased over this century (Cramp, Bourne, and Saunders 1974), numbers of kittiwakes, lesser black-backed gulls, and herring gulls breeding in many Shetland colonies have decreased over the last ten years, while numbers of great skuas have almost stopped increasing. Northern fulmar and northern gannet numbers have continued to increase, while trends in great black-backed gull numbers are uncertain. These trends generally reflect the competitive abilities of these species at fishing boats. It is tempting to suggest that the changes in numbers may be due to the consequences of competitive abilities of these species at fishing boats but it would be unwise to assert this until more is known about the importance of discards and offal in determining the survival rates and breeding success of scavenging seabirds. However, in a parallel situation at garbage dumps in New Jersey, Burger (1981) found that laughing gulls *(larus atricilla)* were unable to successfully compete with herring gulls, and she speculated that this competitive disadvantage prevented the laughing gull population from increasing where herring gulls were abundant and exploited the dumps to their full potential.

It is proposed that the net mesh size used by whitefish trawlers around the British Isles should be increased to 90 millimeters from January 1987, and possibly further increased at successive stages in the future, in order to reduce fishing effort on smaller-size classes of demersal fish. This would have important and interesting implications for scavenging seabirds, since we have shown tht the smaller species cannot utilize the largest discard sizes effectively. An increase in net mesh size will presumably have two immediate effects. First, it will reduce the total quantity of fish discarded. Second, it will increase the mean size of discarded fish. Both of these trends seem likely to reduce food availability to scavenging seabirds, but particularly to herring gulls, lesser black-backed gulls, and great skuas, the species that presently make most use of the smallest discards. We might expect a shift in the balance between species, with northern gannets and great black-backed gulls little affected by the changes but the smaller species finding it increasingly difficult to compete with the larger birds for discarded fish.

■
Summary

Scavenging seabirds can obtain food in the form of fish offal and discarded whole fish from the activities of fishing fleets around the British Isles. The amounts of food made available depend on the fishing method. Whitefish and Norway lobster boats provide most of the food available to scavenging birds at sea. Most discarded whole fish are small (10–25 cm) gadoids from Norway lobster boats or 25–30-centimeter gadoids from whitefish boats. Quantities of offal discarded represent about 11 percent of the mass of gadoids and 6.5 percent of the mass of flatfish processed for market. Crude calculations suggest that offal and discards around the British Isles could support up to 2.5 million 1,000-gram seabirds. Northern fulmars predominate at fishing boats in north and northwest Britain and offshore, with herring gulls predominant further south. Great black-backed gulls are also numerous at fishing boats inshore in the north and northwest, and kittiwakes mainly attend boats further south or east. Northern gannets, great skuas, and lesser black-backed gulls also exploit fishery waste. Adult scavenging seabirds are more efficient at feeding than are juveniles or immature birds. Northern fulmars dominated other species and so obtained almost all of the offal in northern Britain. Northern gannets had a higher foraging success on discards than did other species, and foraging success tended to decline with decreasing body size.

Handling time increased with fish size, and larger seabirds had shorter handling times for a given length of fish. Larger species tended to steal fish from smaller species, although northern gannets were not kleptoparasitic. Flatfish and fish difficult to swallow due to bony exteriors were generally avoided. We estimated that 90 percent of offal and about 75 percent of discarded whole fish were consumed by seabirds. The implications of competitive interactions for food at fishing boats, of net mesh size, and of discarding practices on seabird population dynamics are briefly discussed.

Acknowledgments

This study was supported by a grant from the Natural Environment Research Council. The assistance of the Department of Agriculture and Fisheries for Scotland and the University Marine Research Station at Millport is gratefully acknowledged. Jim Atkinson, Roger Bailey, John R. G. Hislop, Geoff Moore, and Stan Jermyn provided advice, logistical assistance, and comments on an earlier draft. Numerous skippers and crew of fishing boats in Shetland, the Clyde, and Mallaig allowed us to make observations while they were working and gave us generous hospitality on board. Cathy McClaggan kindly assisted with bomb calorimetry. We are also grateful to Joanna Burger and Dick Brown for helpful comments on the submitted manuscript, and thank Dick Brown and Tom Wassenberg for providing useful comparisons with our work from the northwest Atlantic and Australia respectively.

References

Abrams, R. W. 1983. Pelagic seabirds and trawl-fisheries in the southern Benguela Current region. *Mar. Ecol. Prog. Ser.*11:151–156.
Anstey, S. 1984. The foraging at fishing boats by wintering Herring Gulls. Thesis, University of St. Andrews, St. Andrew, Scotland.
Bailey, R. S. and J. R. G. Hislop, 1978. The effects of fisheries on seabirds in the northeast Atlantic. *Ibis* 120:104–105.
Blake, B. F., M. L. Tasker, P. Hope Jones, T. J. Dixon, R. Mitchell and D. R. Langslow. 1984. Seabird distribution in the North Sea. Huntingdon: Nature Conservancy Council. Huntingdon, England.
Boswall, J. 1960. Observations on the use by seabirds of human fishing activities. *British Birds* 53:212–215.
Boswall, J. 1977. The use by seabirds of human fishing activities. *British Birds* 70:79–81.
Brockman, H. J. and C. J. Barnard. 1979. Kleptoparasitism in birds. *Anim. Behav.* 27:487–514.
Brown, R. G. B. 1970. Fulmar distribution: a Canadian perspective. *Ibis* 112:44–51.
Burger, J. 1981. Feeding competition between Laughing Gulls and Herring Gulls at a sanitary landfill. *Condor* 83:328–335.
Burger, J. and Gochfeld, M. 1981. Age-related differences in piracy behaviour of four species of gulls, *Larus. Behaviour* 77:242–67.
Coulson, J. C., P. Monaghan, J. E. L. Butterfield, N. Duncan, K. Ensor, C. Shedden, and C. S. Thomas. 1984. Scandinavian Herring Gulls wintering in Britain. *Ornis Scand.* 15:79–88.
Cramp, S., W. R. P. Bourne, and D. Saunders. 1974. *The Seabirds of Britain and Ireland.* London: Collins.

Ellis, H. 1984. Energetics of free-ranging seabirds. In G. C. Whittow and H. Rahn, eds., *Seabird Energetics*, pp. 203–234. New York: Plenum Press.

Evans, P. G. H. 1984. Status and conservation of seabirds in northwest Europe (excluding Norway and the USSR). In J. P. Croxall, P. G. H. Evans and R. W. Scheiber, eds., *Status and Conservation of the World's Seabirds*, pp. 293–321. International Council for Bird Preservation Cambridge:

Fisher, J. 1952. *The Fulmar*. London: Collins.

Fisher, J. 1966. The Fulmar population of Britain and Ireland, 1959. *Bird Study* 13:5–76.

Furness, R. W. 1982. Competition between fisheries and seabird communities. *Adv. Mar. Biol.* 20:225–307.

Furness, R. W. 1984. Seabird-fisheries relationships in the northeast Atlantic and North Sea. In D. N. Nettleship, G. A. Sauger, and P. F. Springer, eds., *Marine Birds: Their Feeding Ecology and Commercial Fisheries Relationships*, pp. 162–169. Dartmouth, Nova Scotia: Canadian Wildlife Service Special Publication.

Furness, R. W. In press. Kleptoparasitism in seabirds. In J. P. Croxall, ed., *Seabirds: Their Feeding Ecology and Role in Marine Ecosystems*. Cambridge: Cambridge University Press.

Furness, R. W. and J. R. G. Hislop. 1981. Diets and feeding ecology of Great Skuas during the breeding season in Shetland. *J. Zool., Lond.* 195:1–23.

Furness, R. W. and P. Monaghan. 1987. *Seabird Ecology*. Glasgow and London: Blackie.

Furness, R. W. and C. M. Todd. 1984. Diets and feeding of Fulmars *(Fulmarus glacialis)* during the breeding season: a comparison between St. Kilda and Shetland colonies. *Ibis* 126:379–387.

Gaston, A. J. 1985. Energy invested in reproduction by Thick-billed Murres *(Uria lomvia)*. *Auk* 102:447–458.

Gauld, J. A., D. W. McKay, and R. S. Bailey, 1986. Current state of the industrial fisheries. *Fishing News* (June 1986), pp. 30–31.

Greig, S., J. C. Coulson, and P. Monaghan. 1983. Age-related differences in foraging success in the Herring Gull *(Larus argentatus)*. *Anim. Behav.* 31:1237–1243.

Harris, M. P. and J. R. G. Hislop. 1978. The food of young puffins, *Fratercula arctica*. *J. Zool., Lond.* 185:213–236.

Hillis, J. P. 1971. Seabirds scavenging at trawlers in Irish waters. *Irish Nat. J.* 17:129–132.

Hillis, J. P. 1973. Seabirds scavenging at the trawler in the Irish Sea. *Irish Nat. J.* 17.416–418.

Hudson, A. V. 1986. The biology of seabirds utilising fishery waste in Shetland. Ph.D. thesis, University of Glasgow, Scotland.

International Council for the Exploration of the Seas Bulletins Statistique des Peches Maritimes. *Cons. Int. Explor. Mer.* (Copenhagen). Annual issues.

Jermyn, A. S. and A. P. Robb. 1981. Review of the Cod, Haddock and Whiting discarded in the North Sea by Scottish fishing vessels for the period 1975–1980. International Council for Exploration of the Seas Mimeo Report, Denmark.

Kendeigh, S. C., V. R. Dolnik, and V. M. Gavrilov. Avian energetics. In J. Pinowski and S. C. Kendeigh, eds., *Granivorous Birds in Ecosystems*, pp. 127–204. Cambridge: Cambridge University Press.

Lake, N. C. H. 1984. A study of the discarding of fish at sea by commercial fishing vessels—its impact and implications. Thesis, Plymouth Polytechnic, Plymouth, England.

Lockley, R. M. and S. Marchant, 1951. A midsummer visit to Rockall. *British Birds* 44:373–383.

Paul, A. A. and D. A. T. Southgate. 1978. McCance and Widdowson's *The Composition of Foods*. 4th ed. London. HMSO.

Rees, E. I. S. 1963. Marine birds in the Gulf of St. Lawrence and Strait of Belle Isle during November. *Can. Field Nat.* 77:98–107.

Rodriguez, L. 1972. Observaciones sobre aves marinas en las pesquerias del atlantico sudafricano. *Ardeola* 16:159–192.

Salomonsen, F. 1965. The geographical variation of the Fulmar *(Fulmarus glacialis)* and the zones of marine environment in the North Atlantic. *Auk* 82:327–355.

Swennen, C. and P. Duiven. 1977. Size of food objects of three fish-eating seabird species: *Uri aalge, Alca torda,* and *Fratercula arctica* (Aves, Alcidae). *Neth. J. Sea Res.* 11:92–98.

Watson, P. S. 1981. Seabird observations from commercial trawlers in the Irish Sea. *British Birds* 74:82–90.

Wynne-Edwards, V. C. 1962. *Animal Dispersion in Relation to Social Behaviour.* Edinburgh: Oliver and Boyd.

8

Interactions Between Seabirds and Fisheries in the North Pacific Ocean

Linda L. Jones • *National Marine Mammal Laboratory*
Northwest and Alaska Fisheries Center
National Marine Fisheries Service
Seattle, Washington
Anthony R. DeGange • *U.S. Fish and Wildlife Service*
Anchorage, Alaska

Interactions between commercial fisheries and seabirds in the northern Pacific Ocean are increasing with rising consumption of fishery products. As fishing expands into remote areas previously not fished, additional populations of seabirds may be affected. Some interactions such as introduction of fish processing wastes into the environment may be beneficial for seabirds, while others such as competition for fish prey and incidental take by fishing nets may have negatively affected seabird population.

There are a number of factors that make it difficult to determine the effects of commercial fisheries on seabirds. In many instances fisheries are developed and expanded to high levels of effort before research on incidental take and other impacts on seabirds or their prey is initiated. Information on the entanglement of seabirds in commercial fishing gear is difficult to obtain since fishermen generally do not record information on incidental catches of nontarget species. There is often neither the space nor the inclination to permit scientists or observers on commercial vessels to document incidental takes, and data collected from research vessels may not be comparable to that collected during commercial operations because of differences in fishing techniques. The remoteness of fishing areas and the lack of interest of countries that consider seabirds unimportant have also hampered the collection of data on seabirds.

Information on the biology, distribution, and abundance of North

Pacific seabird populations needed for assessment of the effects is frequently lacking. Breeding seabird populations are reasonably well known for Alaska (Sowls et al. 1978), the western continental United States (Sowls et al. 1980; Speich and Wahl in press), British Columbia (Canadian Wildlife Service, unpubl. data), Hawaii (Harrison et al. 1983), and western Mexico (Velarde, unpubl. data). Information on populations offshore in the North Pacific Ocean is more limited, particularly in the western regions. However, faunal descriptions, estimates of densities, and crude population estimates are available for some North Pacific areas (Ainley 1976; Au and Pitman 1986; Bartonek and Gibson 1972; Gould et al. 1982; Harrison et al. 1983; Hunt et al. 1981; Kuroda 1955, 1960; Sanger 1972; Wahl et al. in press; Wiens et al. 1978; Wiens and Scott 1975). Seabird populations in the North Pacific are large, numbering in the low hundreds of millions of individuals, including millions of seasonally migrating shearwaters from the Southern Hemisphere.

In this paper we primarily review interactions between seabirds and commercial fisheries in the North Pacific Ocean and Bering Sea. Reference to interactions in other geographic areas is included where applicable. To make reading easier, scientific names of seabirds are listed with the common names in the appendix.

■
Beneficial Interactions

Although seabird-fisheries interactions are typically viewed as having negative effects on seabird populations, such associations can be benign or even beneficial. For example, fishermen have historically used seabirds as an aid to locating fish schools. In the purse seine fishery of the eastern tropical Pacific (ETP), fishermen use bird flocks to locate schools of tuna (Au and Pitman 1986). Up to 53 percent of bird flocks in the northern ETP were associated with dolphin and tuna schools. The bird species most frequently involved were boobies (Sula spp.), wedge-tailed shearwaters, and jaegers (Stercorarius spp.). Since larger tuna are more frequently associated with the bird flocks, the fishing industry has capitalized on the association.

The availability of fish-processing wastes and garbage benefits surface feeding and scavenging seabirds. Species that have been observed feeding on fish wastes in the northern Pacific include three species of albatross, the northern fulmar, several species of shearwaters (*Puffinus* spp.), the fork-tailed storm petrel, several species of gulls (*Larus* and *Rissa* spp.), and at least two jaegers (De-Gange, Forsell, and Jones 1985). It is not uncommon to see large aggregations of birds in the vicinity of fish-processing vessels, longline vessels, and trawlers. Over 30,000 northern fulmars were estimated by one of the authors (A.R.D.) to be in the vicinity of one large groundfish-processing vessel in the Bering Sea during the summer of 1978. A similar observation was made near longline fishing vessels in the Gulf of Alaska (R. Rowlett, personal communication).

The frequency with which large concentrations of seabirds are encountered feeding on fish-processing wastes and discarded and entangled fish suggests that these food resources are important to seabirds, at least seasonally. Fisher (1966) and Furness (1982) speculate that the availability of fish offal has resulted in increases in populations of northern fulmars and greater black-backed gulls in the northern Atlantic.

In the northwestern Pacific Ocean, fulmars, albatross, shearwaters, kittiwakes, and other seabirds feed on fish and other organisms caught near the surface in offshore gillnets (DeGange, Forsell, and Jones 1985). The nets provide food for the birds; however, there are entanglements of seabirds in the nets while feeding, and fish damage occurs, lowering the market value.

An indirect effect of commercial fisheries is the reduction of populations of seabird competitors. Springer et al. (1986) suggest that in Alaska the removal of walleye pollack *(Theragra chalcogramma)*, which compete with least auklets for copepods, results in increased growth of auklet chicks. Similarly, Furness (1982, 1984) presents evidence that increases in seabird populations in the North Sea after 1900 were in response to the increased abundance of sand lance *(Ammodytes spp.)* and other prey fish species after the intense harvest of competitors such as whiting *(Merlangius mertangus)* and Atlantic cod *(Gadus morhua)*. Similarly, overfishing of Atlantic herring *(Clupea harengus)* and Atlantic mackerel *(Scomber scombrus)* in the North Sea may have resulted in increased populations of sand lance *(Ammodytes maritimus)*, sprats *(Sprattus sprattus)*, and Norway pout *(Trisopterus esmarkii)*, which are important seabird prey (Anderson and Ursin 1977).

■

Competition Between Fisheries and Seabirds

Competition between seabirds and fisheries for prey resources in the northeastern Pacific Ocean has been documented; however, the impact on seabird populations is generally poorly understood.

Competition between seabirds and anchovy *(Engraulis mordax)* fisheries occurs along the western coast of North America from Mexico to Oregon. Wiens and Scott (1975) estimated that off Oregon, sooty shearwaters, Brandt's cormorants, common murres, and Leach's storm-petrel consume 28,000 metric tons of anchovy annually, or as much as 22 percent of the annual anchovy production in the neritic zone. These birds ate more than four times the commercial anchovy catch from Point Conception, California, to the Oregon border. Off southern California, brown pelicans consume about 1 percent of the anchovy biomass annually (Anderson, Gress, and Mais 1982; Anderson and Gress 1984).

Ford et al. (1982) looked at the relationship between prey availability and population status of the common murre in the Bering Sea using a simulation model. They concluded that a decrease of 10–30 percent in prey abundance would cause the murre population to decline, while a decrease of 40 percent would result in complete reproductive failure. If seabird populations are sensitive to prey abundance, then fishery quotas have the potential to seriously impact seabird and other nontarget species.

The crash of the Pacific sardine *(Sardinops sagax)* population may have affected the brown pelican off California. Baldridge (1973) correlated the reproductive success of brown pelicans at Point Lobos, California, with sardine biomass and catch. Pelican breeding virtually ceased at this northernmost nesting site when the sardine population declined.

The breeding success and winter abundance of brown pelicans have also been related to anchovy abundance. A decrease in these parameters, observed as early as 1979 on Los Coronados Islands, Baja California, Mexico, may have been related to the unrestricted anchovy fishery in these waters (Anderson and Gress 1984). Anderson and Gress predict similar negative effects on southern California pelican populations if anchovy harvest levels were increased in that region. The breeding success and abundance of western gulls and Xantus' murrelets similarly rely on anchovy in the Southern California Bight (Hunt and Butler 1980) and could be affected by large commercial harvests.

In British Columbia, overfishing of herring *(C. harengus pallasi)* during the 1950s and 1960s may have resulted in reduced populations of seabirds. Robertson (1972) calculated that herring accounted for 45–100 percent of the diet of glaucous-winged gulls, common guillemots, Brandt's cormorants, Pacific loons, and western grebes. He estimated annual herring consumption to be about 4–7 percent of the adult herring stock in the Gulf of British Columbia, and suggested that this is low because the bird populations have not yet responded to the recovery of the herring stocks.

An indirect effect of competition between fisheries and seabirds is the replacement of a prey species by an unsuitable species. Ainley and Lewis (1974) relate the failure of double-crested cormorants and tufted puffins in central California to recover from low population levels in the early part of this century to the decline of the Pacific sardine as a result of overfishing and climatic changes in the 1940s. These authors suggest that the larger bird species (tufted puffin and double-crested cormorant) relied on the larger sardine, and that its decline either gave competitive advantage to smaller avian predators or decreased the suitability of the region for the larger birds.

■

Incidental Catch

Gillnet fisheries are the only fisheries that commonly incidentally catch seabirds. Gillnets vary in structure depending on target species and location. Surface nets are used for squid and salmon, while bottom nets are used for halibut and croaker.

High Seas Gillnet Fisheries

There are currently five major high seas driftnet fisheries operating in the north Pacific Ocean: Japanese salmon mothership; Japanese land-based salmon; Japanese squid; Republic of Korea squid; and Taiwanese squid (table 8.1). Little information is available concerning the incidental catch of seabirds in the Japanese land-based salmon, Korean squid, and Taiwanese squid gillnet fisheries. Some observations have been made in the Japanese commercial squid fishery. Data on seabird catch by the Japanese salmon mothership fishery have been collected since 1981 by U.S. scientific observers

Table 8.1. High seas gillnet fisheries operating in the North Pacific Ocean.

Fishery	Fishing Season	Net Length (km)	Mesh Size (mm)	Net Depth (m)	No. of Vessels	Fishing Effort[a]
Salmon						
Japanese land-based	Mid-May–mid-July	15	> 110	8	209	8,000–9,000
Japanese mothership	June 1–July 31	15	121,130	8	172	7,000–8,000
Squid						
Japan	June–Dec.	20–50	110–128	6–12	504[b]	32,645[b]
Republic of Korea	April–Jan.	38	96–115	8.8–12.3	97	18,406[c]
Taiwan	May–Dec.	12–25	94	6.5	114	11,389[b]

[a]Number of fishing operations per year. For salmon fisheries, values are the range for the period 1980–1985.
[b]Effort data for 1984 for Japanese squid fishery from report submitted to International North Pacific Fisheries Commission by the Fisheries Agency of Japan, catch and effort statistics for the Japanese squid driftnet fishery in the North Pacific in 1984 (Doc.3111). Taiwanese effort data from unpublished report of Institute of Fishery Biology of Ministry of Economic Affairs and National Taiwan University, Statistics of deep sea jigging and gill net fishery of squid in northwest Pacific region. 1984 fishing season. April 1985.
[c]Effort data for 1985 from Report of the 1986 United States-Republic of Korea Bilateral Meeting on Assessment of North Pacific Fisheries Resources.

on board commercial vessels. The mothership fishery is composed of processing vessels supported by fleets of catcherboats.

Japanese Salmon Fisheries

The Japanese high seas salmon fisheries have operated since about 1952; however, the fishing areas have changed under international agreements. Initially, the high seas salmon fisheries operated throughout the western North Pacific and Bering Sea to 175° W longitude. In 1978, the fishing area was restricted under international agreement (figure 8.1) and fishing effort declined from about 35,000 gillnet operations per year (525,000 km per year) in the land-based and mothership fisheries to 17,000 (255,000 km per year). From 1978 to 1986, there were 172 catcherboats and 4 motherships in the mothership fishery and 209 medium-class vessels in the land-based fishery. As a result of decreased salmon quotas, the number of motherships was reduced in 1987 to 3, with 129 catcherboats. The land-based fishery was reduced to 157 vessels. Fishing effort by these two fisheries declined to 7,391 gillnet operations.

Monofilament gillnets, 15 kilometers in length with stretch mesh sizes from 110–131 millimeters, are set about dusk, drift throughout the night, and are retrieved at dawn. Nets are set at the same

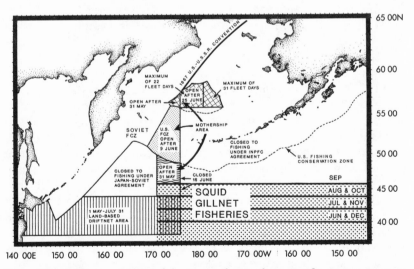

Figure 8.1. High seas driftnet fisheries in the northern Pacific Ocean: Japanese salmon mothership; Japanese salmon land-based; Japanese squid; Korean and Taiwanese Squid. Squid fisheries operate from 170°E to 145°W longitude with the northern boundary changing by month between 40°N and 46°N latitude.

orientation by all vessels (i.e., in parallel rows); however, the orientation varies with geographic areas. Nets extend from the surface to a depth of about 8 meters. The fishing season is from May to early July in the land-based fishery and from June to the end of July in the mothership fishery.

King, Brown, and Sanger (1979) provided estimates of the annual seabird mortality for the Japanese mothership fishery prior to its reduction in size and area in 1978. Using data obtained from twenty research cruises in the mothership fishery area from April to September 1974, they estimated an annual mortality of 75,000 to 250,000 birds by a fleet of 369 catcherboats. They estimated that the combined mortality for the mothership, land-based, and coastal salmon fisheries was 214,500 to 715,000 birds annually. King, Brown, and Sanger (1979) noted that using the same kill rate for the extensive area covered by the three fisheries might not result in a reliable estimate of mortality, and our subsequent research has verified the areal differences in catch rates (DeGange, Forsell, and Jones 1985).

Ogi (1984) estimated that 128,000 to 188,000 seabirds were taken annually in the salmon mothership fishery from 1977 to 1981 based on data collected from Japanese salmon research vessels. The predominant species in the catch were short-tailed shearwaters (46–70 percent) and tufted puffins (14–34 percent). Sooty shearwaters, horned puffins, murres, and other alcids were also observed. Ogi (1984) suggests that the similarity between feeding habits and distribution of short-tailed shearwaters and salmon may account for the large incidental take of this species.

Ainley et al. (1981) estimated that 205,000 birds were killed annually by the mothership salmon fishery based on data from research gillnet operations observed during the 1979 fishing season in the fishing area. The estimate was adjusted upward to 266,500 to account for the lower take rate in research gillnets which use a wide range of mesh sizes besides those used in commercial nets. Fifteen species were observed entangled in gillnets. Of these, the majority (thirteen species) dive or pursuit plunge for their prey while the remainder feed at the surface. Latitudinal differences in the catch rates were also observed by Ainley et al. (1981). Rates increased logarithmically as fishing approached the Aleutian Islands. Near the islands, entangled birds were mainly breeding adults, whereas farther offshore immature birds dominated.

The most extensive data on seabird mortality in the Japanese salmon mothership fishery were collected by U.S. observers on commercial catcherboats from 1981 to 1984 (DeGange, Forsell, and

Jones 1985). Approximately 6 percent of the gillnet retrievals were monitored each year in the U.S. Exclusive Economic Zone (EEZ). The distribution of observations was reflective of the fishing effort.

For the period 1982–84, over 24,000 seabirds were observed in the nets. The dominant species was short-tailed shearwaters (figure 8.2). Up to 6 percent were released alive each year. Alcids and

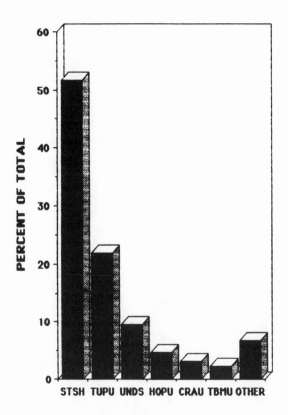

SPECIES COMPOSITION OF KILL, 1981-1984

SPECIES

Figure 8.2. Species composition of seabirds killed in the Japanese salmon mothership fishery, 1981–84. STSH = short-tailed shearwater; TUPU = tufted puffin; UNDS = unidentified dark shearwater; HOPU = horned puffin; CRAU = crested auklet; TBMU = thick-billed murre; other = common murre, northern fulmar, Laysan albatross, fork-tailed storm petrel, and ten other species.

Table 8.2. Estimates of total mortality of seabirds in the Japanese salmon mothership fishery by species and year, 1981–1984 (from DeGange, Forsell, and Jones 1985).

Species	1981	1982	1983	1984	$\bar{x}^a$
Laysan albatross	228	0	0	114	86
Northern fulmar	3,398	1,682	2,846	1,483	2,352
Scoty shearwater	62	2,164	370	399	749
Short-tailed shearwater	9,901	67,597	176,069	60,977	78,636
Unidentified shearwater	60,108	2,672	1,584	0	16,091
Fork-tailed storm petrel	1,350	828	1,014	626	954
Leach's storm-petrel	194	0	123	57	94
Pomarine jaeger	25	25	0	13	16
Black-legged kittiwake	153	165	62	0	95
Dovekie	0	1	0	0	>1
Common murre	1,474	1,352	1,755	1,004	1,396
Thick-billed murre	651	4,715	12,726	2,729	5,205
Unidentified murre	2,388	6,810	239	97	2,205
Pigeon guillemot	25	0	25	0	13
Ancient murrelet	415	2,323	2,428	968	1,533
Cassin's auklet	249	290	618	57	304
Parakeet auklet	290	688	830	525	583
Least auklet	25	290	357	207	219
Crested auklet	12,528	2,228	7,219	285	5,565
Unidentified small alcid	1,245	352	364	224	546
Rhinoceros auklet	50	25	0	0	19
Tufted puffin	26,302	73,192	35,328	19,581	38,600
Horned puffin	5,479	13,187	7,222	7,108	8,249

Unidentified puffin	2,367	0	0	0	592
Unidentified bird	877	290	62	0	307
Total ±	129,784 ±	180,875 ±	251,426 ±	96,397 ±	164,621
95% confidence interval[b]	122,628	63,297	80,181	27,670	

[a]Mean calculated by averaging across four years.

[b]We have data for only two sets from the northern U.S. Exclusive Economic Zone (EEZ) in 1981; consequently 95 percent confidence intervals around the mean number of birds killed/set for two species (short-tailed shearwater and crested auklet) were extremely wide. For the northern U.S. EEZ in 1981 alone, the estimated number of birds killed was 18,063 ± 84,197.

fulmars were generally in good condition upon release, while half of the shearwaters may not have survived.

Most gillnets caught only a few birds (< 20). However, up to 457 birds were recorded in a single retrieval (15 km of net). Catch rates varied between years, with 1983 having the highest rate, 1984 the lowest. The mean number of birds per gillnet ranged from 11.6 to 34.1 annually.

Degange, Forsell, and Jones (1985) reported annual and spatial variation in the catch rate of seabirds. For most species, catches were highest on the Continental Shelf, the shelf break, and in the Alaska Current. The lowest catch rates were in the Bering Sea, while the highest rates were south of the Aleutians. The catch of breeding birds was highest near the Aleutian Islands.

Incidental catch rates varied over the fishing season for some species. Short-tailed shearwaters showed higher rates in mid- to late June and again in mid- to late July. Tufted puffins also had a peak in entanglement rates in mid- to late June, and in some years another peak in late July.

Based on the observed commercial operations, annual mortalities were estimated for each species of seabird (table 8.2). The total estimated annual bird mortality in the salmon mothership fishery ranged from about 96,000 (1984) to over 250,000 (1983) (DeGange, Forsell, and Jones 1985). These estimates do not include dropouts, birds released alive, or take by the land-based salmon fishery.

Sano (1978) estimated the incidental take in the land-based salmon fishery to be at least 113,000 to 232,000 seabirds annually. We estimated the annual mortality in the land-based fishery based on 296 observations during Japanese salmon research vessel gillnet sets in the land-based fishing area in 1977 (table 8.3) to be approximately 124,700 seabirds annually.

Squid Gillnet Fisheries

Squid fisheries began in the North Pacific Ocean in the late 1970s and expanded rapidly. These fisheries extend from off Japan to longitude 145°W (figure 8.1) and occur from May to December. The type of gear and fishing methods are similar to the salmon fisheries. Gillnets range in length from 20–50 kilometers, go to depths of 12 meters, and have mesh sizes of 80–120 millimeters. Nets are set at dusk in an east-west orientation and retrieved at daybreak. The total number of squid boats in operation in 1984 was about 700 and the total fishing effort was over 50,000 gillnet operations, or over

Table 8.3. Incidental take of seabirds by high seas Japanese land-based salmon and squid fisheries in the northern Pacific Ocean.

Species	LAND-BASED SALMON FISHERY[a]	
	Percent in catch[a]	Estimated Annual Take
Short-tailed shearwater	30.6	38,210
Unidentified shearwater	21.7	27,115
Tufted puffin	20.9	26,090
Thick-billed murre	11.4	14,175
Parakeet auklet	4.7	5,958
Rhinoceros auklet	3.3	4,108
Common murre	1.5	1,850
Sooty shearwater	1.3	1,643
Fork-tailed storm petrel	1.3	1,643
Northern fulmar	1.1	1,438
Horned puffin	0.8	1,025
Laysan albatross	0.7	822
Pigeon guillemot	0.3	410
Crested auklet	0.1	—
Pomarine jaeger	0.1	—
TOTAL		124,479

	HIGH SEAS SQUID GILLNET FISHERY[b]	
	Percent in catch	Estimated Annual Take[c]
Unidentified shearwaters	63.6	N.A.
Black-footed albatross	14.6	N.A.
Laysan albatross	10.9	N.A.
Flesh-footed shearwater	3.6	N.A.
Sooty shearwater	3.6	N.A.
Buller's shearwater	1.8	N.A.
Short-tailed shearwater	1.8	N.A.

[a] From DeGange, Forsell, and Jones (1985). Data collected in 1977.
[b] L. Tsunoda, unpublished data from thirty observed gillnet operations in 1986 aboard a Japanese commercial squid vessel fishing in the northeastern area of the squid fishery.
[c] N.A. = data not available.

1.5 million kilometers of gillnet. The target species is the flying squid *(Ommastrephes bartrami)*.

Data were collected on seabird entanglements in the Japanese squid fishery during 30 gillnet retrievals in July–August 1986, in the eastern North Pacific between 40°53' and 43°56'N latitude and 158°14' and 169°57'W longitude. A total of fifty-three seabirds were caught: two flesh-footed shearwaters, thirty-six short-tailed or sooty shearwaters, six Laysan albatross, eight black-footed albatross, and one New Zealand shearwater (table 8.3).

In 1985, eight gillnet retrievals (a total of 53 nm of gillnet) were monitored from a motor launch near Japanese commercial squid vessels. Observations were made from August 15 to September 19 in two areas: near 44°N latitude and 153°W longitude; and near 46°N latitude and 148°W longitude. Seabirds were infrequently caught, with a total of twenty-seven shearwaters and one black-footed albatross observed in the nets (Ignell, Bailey, and Joyce 1986).

Numbers of individuals as well as the number of species of seabirds observed in squid gillnets in the northeastern Pacific are substantially less than is the case for the Japanese salmon mothership gillnet fishery (table 8.3). However, the number of gillnet operations observed has been relatively small and observations were from a limited area of the fishery.

Coastal Gillnet Fisheries

The Salmon Fishery

In British Columbia, Canada, a salmon fishery operates from February to October in Barkley Sound, off southwestern Vancouver Island (Carter and Sealy 1984). This fishery involves small boats (about 10 m in length) using gillnets throughout the day and night, depending on the fishery location, and returning to port daily to offload. The gillnets are monofilament, 135–375 meters in length, extend from the surface to a depth of 10 meters, and are usually set perpendicular to the shore. Mesh size is 100 to 130 millimeters (Carter and Sealy 1984).

The predominant species incidentally taken in this fishery was marbled murrelets; common murres and rhinoceros auklets were also killed (Carter and Sealy 1984). The majority of the murrelets were in breeding condition and were killed at night, within 2 meters of the surface. Murres were caught both day and night.

Carter and Sealy (1984) estimated that 175 to 250 marbled murrelets were killed in 1980 by the coastal salmon fishery. The authors estimated that 6 percent of the local breeding murrelets and, in consequence, 13 percent of the young that were not raised to fledglings were killed in that year. Since the mortality occurs annually in a local population, there could be a long-term effect from the fishery. Carter and Sealy (1984) noted that the catch level was related to the amount of fishing activity, weather conditions, and density of the birds, as well as their behavior. As Ainley et al. (1981) also pointed out, alcids are susceptible to capture by gillnets because of their diving behavior. Carter and Sealy (1984) observed small flocks of murrelets near the nets and suggested the birds may be attracted to small fish aggregating near the nets. M. E. Islieb (personal communication, in Carter and Sealy 1984) reported a similar attraction to gillnets by marbled murrelets in Alaska.

The California White Croaker Fishery

Off California, the white croaker (Genyonemus lineatus) fishery operates year round; however, regulations enacted in 1984 closed areas less than 40 meters deep to gillnet and trammel fishing from May to September. This regulation significantly reduced fishing efforts during those months. The fishery operates north of Half Moon Bay and off Bolinas Bay in central California, with vessels returning to port daily. The monofilament gillnets are 1 to 3 kilometers in length, extend from the surface to 3 to 4 meters depth, and have a mesh size of 60 to 76 millimeters. Nets are anchored and fished from 1 to 22 hours, but usually less than 7 hours (Wild 1986).

Of thirty-four net retrievals observed in the white croaker fishery in 1984, fifteen had seabirds entangled. The catch included 84 common murres, 1 pelagic cormorant, and 1 western grebe. The observed catch in 1983 under the previous regulations was 152 common murres in thirteen retrievals (Wild 1986).

The California Halibut Fishery

The California halibut (Paralichthys californicus) fishery operates off central California from May to October, with the most activity in July and August (Wild 1986). Gillnets, suspendered gillnets, and trammel nets are used in water depths up to 16 meters. In 1983, water depths regulations limited fishing to water depths deeper than 20 meters (Heneman 1983). Legislation passed in 1987 to protect marine birds and mammals bans all gillnets in state waters north of Point Reyes and restricts gillnets to waters deeper than 40 meters from Point Reyes south to Monterey Bay. Nets are anchored and are usually set for less than one day but can be set for up to four days. Net length initially was 100 to 600 meters, consisting of 100-meter shackles. In 1984, the amount of net was regulated to 100 meters or less. Mesh size is 200–250 millimeters; net depth is 3 to 6 meters (Wild 1986).

In 1981, the California Department of Fish and Game estimated that 20,000 seabirds were taken annually by the halibut fishery (Heneman 1983). Species caught were common murres, pelagic and Brandt's cormorants, pigeon guillemots, white-winged and surf sco-ters, common loons, and western grebes. In 1983, 118 gillnet re-trievals were monitored from April to July. Over 700 seabirds were caught (Heneman 1983). The mortality estimate for the dominant species was 25,000 to 30,000 common murres, or more than 19 percent of the breeding population of central California (Heneman 1983). The mortality was highest in July (9 birds/gillnet set). With the 1984 regulations which restricted fishing to deeper waters, the estimated mortality dropped to 6,000 to 8,000 common murres in 1984 and 7,000 to 10,000 in 1985 (P. Wild, Calif. Dept. of Fish and Game, personal communication). The catch rates were 42 birds per 1,000 fathoms of net in 1983 and 9 in 1984 (Wild 1986).

The California Shark Fishery

Low numbers of seabirds have been taken in shark gillnets off central and northern California (Wild 1986). Nets are of two types: nylon twine gillnets, 334 meters in length, 3 meters in depth, with 152-millimeter mesh; and suspendered monofilament gillnets, 400 meters in length, 3 meters in depth, with 406-millimeter mesh. Nets are set at the surface or on the bottom for about twenty-four hours in water depths of 34 to 70 meters. Various shark species are caught. Fifteen common murres were entangled in fifteen observed fishing operations in 1984, a take rate of six birds per thousand fathoms of net.

■
Net Debris

Lost, abandoned, and discarded gillnet material is another source of mortality for seabirds. DeGange and Newby (1980) reported ninety-nine seabirds of five species entangled in 1,500 meters of lost salmon driftnet in the western North Pacific. An abandoned salmon gillnet was retrieved near Agattu Island in 1979 with more than four hundred seabirds entangled (L. Jones, unpublished data). The amount of gill-net debris and its effective fishing time is unknown but it potentially could entangle large numbers of seabirds.

■
Discussion

The beneficial effects of fish wastes produced by fishing vessels have been noted in the North Pacific as well as other areas (Fisher 1966; Furness 1982). However, as domestic vessels replace foreign vessels in the U.S. EEZ, this benefit may decrease. Previously, fish processing was done throughout the fishing areas and the year by a large number of vessels. Although the volume of materials produced will remain the same, the time period and area where processing is done is decreasing (Loh-Lee Low, National Marine Fisheries Service, personal communication, 1987) resulting in the concentration of these food resources in time and space. How the seabird populations will be affected will be a function of their dependency on these resources and the availability of other prey.

Seabirds are likely to be affected by intense competition from fisheries for several reasons: many seabirds utilize fish resources and have limited capability to exploit different resources; breeding seabirds require sufficient prey within about 50 kilometers of the nesting area to support both chicks and parents; and some seabird species are restricted to shallow foraging depths, experiencing reproductive failures when prey resources are depleted in these waters.

The effects of competition between seabirds and fisheries for the same prey resources have been documented for several species in the North Pacific (Ainley and Lewis 1974; Anderson, Gress, and Mais 1982; Anderson and Gress 1984; Baldridge 1973; Robertson 1972; Springer et al. 1986; Wiens and Scott 1975). The development or expansion of fisheries that compete with seabirds for prey resources potentially will affect additional seabird populations. For example, the expanding fisheries off California for squid and anchovy will compete with shearwaters (Shallenberger 1984), and in the Gulf of Alaska and Bering Sea the development of fisheries for capelin *(Mallotus villosus)* and Pacific sand lance *(Ammodytes hexapterus)* will result in competition with a number of seabird species (DeGange and Sanger 1987; Sanger in press). To protect seabird populations, particularly nesting populations, sufficient prey resources for the bird populations need to be considered in resource allocations. As Anderson and Gress (1984) point out, such management becomes more difficult when prey resources cross international boundaries. If anchovy stocks were to be overharvested in Mexico, stocks off California could decline, which would have an impact on nesting brown pelican populations in the United States.

The effects of gillnet fisheries on seabird populations depend on fishing techniques and the location and time period of the fishery. Along the California coast, gillnet fishing reduced populations of common murres and other seabird species until regulations limiting water depths of fishing were introduced (Carter 1986). Additional measures including area closures may be required to fully protect the bird populations.

The closure of a 60-nautical-mile area around the Aleutian Islands has been proposed to protect breeding seabird colonies from the high seas gillnet fisheries. This restriction would not substantially reduce the overall take of seabirds because additional fishing effort could occur in high-density areas outside the closed area. However, the composition of the take would probably shift from breeding to nonbreeding birds, spreading the impact over a greater number of colonies and having a smaller impact on the populations.

The effects of the high seas gillnet fisheries on seabird populations are not clear. Most of the alcids taken in the land-based and mothership salmon fisheries are nonbreeding and immature birds, originating from many colonies in the western North Pacific and Aleutian Islands. Thus, impacts on seabird populations are likely spread over many colonies. Lack of historical data, particularly from the western Aleutian Islands, has compromised our ability to assess changes in the sizes of breeding populations as a result of the gillnet takes.

Although there is information on the incidental catch of seabirds by the land-based salmon fishery (Sano 1978; Ogi 1984; DeGange, Forsell, and Jones 1985) and the mothership fishery (DeGange, Forsell, and Jones 1985), minimal information has been obtained on the squid gillnet fisheries. Catch rates may be lower in the squid fisheries (Ignell, Bailey, and Joyce 1986; DeGange, Forsell, and Jones 1985; L. Tsunoda, unpublished data); however, the temporal and spatial magnitude of this fishery suggests that large numbers of seabirds could be taken. Without more information on the magnitude and age and sex composition of the take, and the abundance and survival rates of the affected populations, the effect of the squid gillnet fisheries on seabird populations can not be determined.

■
Summary

Interactions between seabirds and commercial fisheries in the North Pacific Ocean ranging from beneficial to detrimental are reviewed. Beneficial interactions include the availability of new food resources, and reductions in abundance of seabird competitors. Competition between seabirds and commercial fisheries for resources such as herring, anchovy, and sardine has affected reproductive success and abundance of seabirds along the west coast of North America. Incidental take of seabirds by gillnet fisheries has been of concern. Estimates of the annual catch by the Japanese high seas and coastal salmon gillnet fisheries have been as high as 750,000 birds, based on data collected on salmon research vessels. Data on the incidental take by the Japanese salmon mothership fishery were collected by U.S. observers from 1981 to 1984. Based on these data, the estimated annual take of seabirds ranged from 96,000 to over 250,000. The incidental take by land-based salmon fishery was estimated to be between 113,000 to 232,000 annually. Insufficient information is available to estimate the catch by the high seas squid fisheries, however the temporal and spatial magnitude of these fisheries suggests that large numbers of seabirds could be taken each year. Coastal gillnet fisheries off North America have also caught seabirds. Legislation has been enacted in California to protect the seabird populations.

Acknowledgments

We would like to thank Paul Wild and Doyle Hanen for their help in providing information for this manuscript; Larry Tsunoda for data from the Japanese squid gillnet fishery; Joanna Burger, Robert Furness, Mary Nerini, Dale Rice, Robert De-Long, Robert Miller, Sally Mizroch, and Thomas Loughlin for reviewing earlier drafts; and Leola Hietela for typing.

References

Ainley, D. G. 1976. The occurrence of seabirds in the coastal region of California. *Western Birds* 7:33–68.

Ainley, D. G. and T. J. Lewis. 1974. The history of Farallon Island marine bird populations, 1854–1972. *Condor* 76:432–446.

Ainley, D. G., A. R. DeGange, L. L. Jones, and R. J. Beach. 1981. Mortality of seabirds in high-seas salmon gillnets. *Fish. Bull.* 79:800–806.

Anderson, D. W. and F. Gress. 1984. Brown pelicans and the anchovy fishery off southern California. In D. N. Nettleship, G. A. Sanger, and P. F. Springer, eds., *Marine Birds: Their Feeding Ecology and Commercial Fisheries Relationships*, pp. 128–135. Dartmouth, Nova Scotia: Canadian Wildlife Service Special Publication.

Anderson, D. W., F. Gress, and U. F. Mais. 1982. Brown pelicans: influence of food supply on reproduction. *Oinos* 39:23–31.

Anderson, K. P. and E. Ursin. 1977. A multispecies extension to the Beverton and Holt theory of fishing, with accounts of phosphorus circulation and primary production. *Medd. Dan. Fisk., og Havunders.* n.s. 7:319–435.

Au, D. W. K. and R. L. Pitman. 1986. Seabird interactions with dolphins and tuna in the eastern tropical Pacific. *Condor* 88:304–317.

Baldridge, A. 1973. The status of the brown pelican in the Monterey region of California: past and present. *Western Birds* 4:93–100.

Bartonek, J. C. and D. D. Gibson. 1972. Summer distribution of pelagic birds in Bristol Bay, Alaska. *Condor* 74:416–422.

Carter, H. R. 1986. Rise and fall of the Farallon common murre. *Pt. Reyes Bird Observatory Newsletter* 72:1–4.

Carter, H. R. and S. G. Sealy. 1984. Marbled murrelet mortality due to gillnet fishery in Barkley Sound, British Columbia. In D. N. Nettleship, G. A. Sanger, and P. F. Springer, eds., *Marine Birds: Their Feeding Ecology and Commercial Fisheries Relationships*, pp. 212–220. Dartmouth, Nova Scotia: Canadian Wildlife Service Special Publication.

DeGange, A. R., D. G. Forsell, and L. L. Jones. 1985. Mortality of seabirds in the high-seas Japanese salmon mothership fishery, 1981–1984. Manuscript. U.S. Fish and Wildlife Service, Anchorage, Alaska.

DeGange, A. R. and T. C. Newby. 1980. Mortality of seabirds and fish in a lost salmon driftnet. *Mar. Pollu. Bull.* 11:322–323.

DeGange, A. R. and G. A. Sanger. 1987. Marine birds of the Gulf of Alaska. In D. W. Hood and S. T. Zimmerman, eds., *The Gulf of Alaska: Physical Environment and Biological Resources*, pp. 479–524. Washington, D.C.: U.S. Department of Commerce and the Interior, Ocean Assessment Division, National Oceanic and Atmospheric Administration.

Fisher, J. 1966. The Fulmar population of Britain and Ireland, 1959. *Bird Study* 13:5–76.

Ford, R. G., J. A. Wiens, D. Heinemann, and G. L. Hunt. 1982. Modeling the sensitivity of colonially breeding marine birds to oil spills: guillemot and kittiwake populations on the Pribilof Islands, Bering Sea. *J. appl. ecol.* 19:1–32.

Furness, R. W. 1982. Competition between fisheries and seabird communities. *Adv. Mar. Biol.* 20:225–307.

Furness, R. W. 1984. Seabird-fisheries relationships in the northeast Atlantic and North Sea. In D. N. Nettleship, G. A. Sanger, and P. F. Springer, eds., *Marine Birds: Their Feeding Ecology and Commercial Fisheries Relationships*. Dartmouth, Nova Scotia: Canadian Wildlife Service Special Publication.

Gould, P. J., D. J. Forsell, and C. J. Lensink. 1982. Pelagic distributions and abundance of seabirds in the Gulf of Alaska and eastern Bering Sea. U.S. Fish and Wildlife Service, Biological Service Program. FWS/OBS-82/48. Anchorage, Alaska.

Harrison, C. S., T. S. Hida, and M. P. Seki. 1983. Hawaiian seabird feeding ecology. *Wildl. Monog.* 85:1–71.

Heneman, B. 1983. Gillnets and seabirds 1983. *Pt. Reyes Bird Observatory Newsletter* 63:1–3.

Hunt Jr., G. L. and J. L. Butler. 1980. Reproductive ecology of western gulls and

Xantus' murrelets with respect to food resources in the Southern California Bight. *CalCOFI Rep.* 21:62–67.

Hunt Jr., G. L., P. J. Gould, D. J. Forsell, and H. Peterson, Jr. 1981. Pelagic distribution of marine birds in the eastern Bering Sea. In D. W. Hood and J. A. Calder, eds., *The Eastern Bering Sea Shelf: Oceanography and Resources*, pp. 689–718. Washington, D.C.: Office of Marine Pollution Assessment, National Oceanic and Atmospheric Administration.

Ignell, S., J. Bailey, and J. Joyce. 1986. Observations on high-seas squid gill-net fisheries, North Pacific Ocean, 1985. National Oceanic and Atmospheric Administration Tech. Memo. Seattle, Washington. National Marine Fisheries Service F/NWC-105.

King, W. B., R. G. B. Brown, and G. A. Sanger. 1979. Mortality to marine birds through commercial fishing. In J. C. Bartonek and D. N. Nettleship, eds., *Conservation of Marine Birds of Northern North America*, pp. 195–199. Washington, D.C.: U.S. Fish and Wildlife Service, Wildlife Research Report no. 11.

Kuroda, N. 1960. Analysis of seabird distribution in the northwest Pacific Ocean. *Pac. Sci.* 14:55–67.

Ogi, H. 1984. Seabird mortality incidental to the Japanese salmon gill-net fishery. In J. P. Croxall, P. G. H. Evans, and R. W. Schreiber, eds., *Status and Conservation of the World's Seabirds*. Norwich, England: Page. (International Council for Bird Preservation Technical Publication no. 2:717–721).

Robertson, I. 1972. Studies on fish-eating birds and their influence on stocks of Pacific herring in the Gulf Islands of British Columbia. Manuscript. Pac. Bio. Sta. (Nanaimo, B.C.)

Sanger, G. A. 1972. Preliminary standing stock and biomass estimates of seabirds in the subarctic Pacific region. In A. Y. Takenouti, ed. *Biological Oceanography of the Northern North Pacific Ocean*, pp. 589–611. Tokyo: Idemitsu-Shoten.

Sanger, G. A. In press. Diets and food web-relationships of seabirds in the Gulf of Alaska and ajacent marine regions. Unpublished report to the Outer Continental Shelf Environmental Assessment Program. Anchorage, Alaska.

Sano, O. 1978. Seabirds entangled in salmon drift nets. *Enyo* 30:1–4.

Shallenberger, R. J. 1984. Fulmars, shearwaters and gadfly petrels: family Procellariidae. In D. Haley, ed., *Seabirds of Eastern North Pacific and Arctic Waters*, pp. 42–56. Seattle, Wa.: Pacific Search Press.

Sowl, L. W. and J. C. Bartonek. 1974. Seabirds—Alaska's most neglected resource. *Trans. 39th N. Amer. Wildl. Nat. Resources Conf.*, pp. 117–126. Washington, D.C.: Wildlife Management Institute.

Sowls, A. L., S. A. Hatch, and C. J. Lensink. 1978. *Catalog of Alaska Seabird Colonies.* Washington, D.C.: U.S. Fish and Wildlife Service. Biological Service Program, FWS/OBS-78/78.

Sowls, A. L., A. R. DeGange, J. W. Nelson, and G. S. Lester. 1980. *Catalog of California Seabird Colonies.* Washington, D.C.: U.S. Fish and Wildlife Service. Biological Service Program, FWS/OBS-80/37.

Speich, S. M. and T. R. Wahl. In press. *Catalog of Washington Seabird Colonies.* Washington, D.C.: U.S. Fish and Wildlife Service, Biological Service Program.

Springer, A. M., D. G. Roseneau, D. S. Lloyd, C. P. McRoy, and E. C. Murphy. 1986. Seabird responses to fluctuating prey availability in the eastern Bering Sea. *Mar. Ecol. Prog. Ser.* 32:1–12.

Wahl, T. R., D. G. Ainley, A. Benedict, and A. R. DeGange. 1988. Associations between seabirds and water masses in the northern Pacific Ocean. Manuscript. Bellingham, Wash.

Wiens, J. A. and J. M. Scott. 1975. Model estimation of energy flow in Oregon coastal seabird populations. *Condor* 77:439–452.

Wiens, J. A., D. Heinemann, and W. Hoffman. 1978. Community structure, distribution and interrelationships of marine birds in the Gulf of Alaska. In *Environmental Assessment of the Alaska Continental Shelf*, Biological Studies 3; 1–178. Boulder, Colo.: National Oceanic and Atmospheric Administration, Resource Laboratory.

Wild, P. W. 1986. Progress report: central California gill and trammel net investigations (northern area) 1984. California Department of Fish and Game, unpublished report. Long Beach, Calif.

Appendix 8.1. Scientific and common names of seabirds referred to in text.

Common Name	Scientific Names
Black-footed albatross	*Diomedea nigripes*
Black-footed kittiwake	*Rissa tridactyla*
Brandt's cormorant	*Phalacrocorax penicillatus*
Brown pelican	*Pelecanus occidentalis*
Common guillemot	*Uria aalge*
Common loon	*Gavia immer*
Common murre	*Uria aalge*
Double-crested cormorant	*Phalacrocorax auritus*
Flesh-footed shearwater	*Puffinus carneipes*
Fork-tailed storm petrel	*Oceanodroma furcata*
Glaucous-winged gull	*Larus glaucescens*
Greater black-backed gull	*Larus marinus*
Horned puffin	*Fratercula corniculata*
Laysan albatross	*Diomedea immutabilis*
Leach's storm petrel	*Oceanodroma leucorhoa*
Least auklet	*Aethia pusilla*
Marbled murrelet	*Brachyramphus marmoratus*
New Zealand shearwater	*Puffinus bulleri*
Pacific loon	*Gavia arctica*
Pelagic cormorant	*Phalacrocorax pelagicus*
Pigeon guillemot	*Cepphus columba*
Rhinoceros auklet	*Cerorhinca monocerata*
Short-tailed albatross	*Diomedea albatrus*
Short-tailed shearwater	*Puffinus tenuirostris*
Sooty shearwater	*Puffinus grisseus*
Surf scoter	*Melanitta perspicillta*
Tufted puffin	*Fratercula cirrhata*
Wedge-tailed shearwater	*Puffinus pacificus*
Western grebe	*Aechmophorus occidentalis*
Western gull	*Larus occidentalis*
White-winged scoter	*Melanitta fusca*
Xantus' murrelet	*Synthliboramphus hypoleucus*

9

■ Interactions Between Seabirds and Marine Mammals Through Time: Interference Competition at Breeding Sites

Kenneth I. Warheit · *Department of Paleontology*
University of California (Berkeley)
David R. Lindberg · *Museum of Paleontology*
University of California (Berkeley)

Long-term ecological studies of animal communities are rare. The time span in which studies are conducted is often constrained by the length of time and amount of funding available to dedicated workers and their intellectual descendants. However, despite such temporal limitations, it is often assumed that long-term ecological studies identify the major selective forces that have shaped the evolution of the population or the structure of the community.

Because seabirds are long lived and their habitat (i.e., the ocean) undergoes long-term cyclic changes (e.g., El Niño events), studies covering many generations are necessary to understand even the most basic mechanisms of population and community structure. Not surprisingly, seabirds, perhaps more than any other group of animals, have been the subject of many long-term ecological studies. In population and community ecology, such studies often range from ten to thirty years (e.g., Ainley, LeResche, and Sladen 1983; Coulson and Thomas 1985; Ainley and Boekelheide in press, among others). A few authors have attempted to expand the temporal boundaries of their research by using prehistoric or historical information about particular populations and/or localities under study. For example, Ainley and Lewis (1974) used the writings of nineteenth- or early twentieth-century naturalists to document population size on Southeast Farallon Island, California as far back as 1854. Similar studies have been conducted for some Indian Ocean bird communities (Diamond 1979; Diamond and Feare 1980). Whereas these studies have clearly contributed to the understanding of historical factors that have structured Recent seabird populations, they have not necessarily helped identify processes that have acted on ancient populations and communities in the not-so-near past. Through the study of fossils we can directly assess at least some aspects of the role of speciation and extinction in determining the structure of modern communities (Fowler and MacMahon 1982; Janzen and Martin 1982).

In this paper we use both the fossil record and observations of Recent marine communities to examine the possible role of interference competition (see Maurer 1984; preemptive or encounter competition in Schoener 1983) between pinnipeds and seabirds in structuring seabird communities. The interference competition that we address is competition for space at breeding sites. We recognize that trophic interactions between marine mammals and seabirds (exploitative competition) could be important (but perhaps from a more commensalistic point of view; see other papers in this volume), but we have chosen to emphasize spatial interactions that occur in seabird breeding habitats. At these sites selection is hard and can operate directly on progeny. We believe that competition for space between these marine animals has been previously overlooked for two reasons. First, both seabird and pinniped populations have been so depressed in historical times by human activities (see Olson and James 1982a, 1982b; Steadman and Olson 1985 for discussion of the effects of aboriginal people on island bird populations) that interactions that may have occurred in the past are rare in the Recent. Second, the large-scale competitive interactions between pinnipeds and seabirds that would influence the community structure may only occur during times of environmental change (e.g., tectonic subsidence or eustatic changes to sea level). These changes occur on a geologic time scale and are not recognizable in an ecological framework.

■

Initial Observations and Hypotheses

Our interest in competition between pinnipeds and seabirds was sparked by the generic diversity pattern of northern Pacific diving marine birds through time (Lindberg and Kellogg 1982) and observations of oystercatchers *(Haematopus)* interacting with pinnipeds at several localities along the California coast. With subsequent work we were able to document that pinniped disturbance at American black oystercatcher *(H. bachmani)* breeding sites could substantially lower the bird's reproductive success (Warheit, Lindberg, and Boekelheide 1984). These spatial interactions are not limited to oystercatchers and pinnipeds, but have also been observed in other seabird rookeries in California (Ainley and Boekelheide in press). Furthermore, Sowls, Hatch, and Lensink (1978) have suggested that a large murre colony on Walrus Island, Pribilof Islands, Alaska may have been displaced by Steller's sea lions *(Eumetopias jubatus)*. In the Southern Hemisphere, Croxall et al. (1984) have shown that petrel and wandering albatross *(Diomedia exulans)* breeding habitats are being destroyed by the expanding population of Antarctic fur seals *(Arctocephalus gazella)* on Bird Island, South Georgia. Also, Rand (1951, 1952) and Shaughnessy (1980, 1984) have concluded that disturbance by South African fur seals *(Arctocephalus pusillus)* was responsible, in part, for the declines in seabird populations (in particular cape gannets *[Morus capensis]* and jackass penguins *[Spheniscus demersus]*) on a number of South African and Namibian islands during this century. These examples, while anecdotal, suggest that these types of interactions are not isolated incidents. We argue that they may represent important structuring phenomena in seabird communities.

Competition for space between seabirds and pinnipeds may arise for two related reasons. First, both pinnipeds and seabirds combine offshore marine feeding with terrestrial breeding. This results in similar requirements for breeding sites: they both need appropriate terrain that has ready access to the oceanic environment, and is free of terrestrial predators. Second, both groups tend to be gregarious breeders (see also Bartholomew [1970] and Stirling [1975]).

Not all pinnipeds are gregarious, and not all breed on land (King 1983a). As a result, only particular groups of pinnipeds are potential competitors for space with nesting seabirds. Gregariousness is associated with polygyny in pinnipeds (Bartholomew 1970). Pinnipeds with polygynous land-based breeding systems include all species in the Otariidae (fur seals and sea lions), the Odobenidae (walruses; walruses pup on land, but copulate in the water [King 1983a; Tagert personal communication]), and a few Phocidae (true seals) such as *Mirounga* (elephant seals) (Bartholomew 1970; King 1983a; Stirling 1975). These polygynous pinnipeds are also sexually dimorphic and therefore their polygyny (and gregarious behavior) can be recognized in the fossil record (Bartholomew 1970; Repenning 1976). The desmatophocids, an extinct otariid group from the middle to late Miocene of California, were sexually dimorphic, which suggests that polygynous and gregarious behavior has occurred in pinnipeds for at least fifteen million years (Mitchell 1965, 1966; Repenning 1976; Barnes 1972).

Figure 9.1 shows how Recent gregarious pinnipeds can occupy available space on sandy and rocky beaches. The degree to which these pinnipeds exclude seabirds from nesting habitats is a function of the availability of alternative habitats and the ability of seabirds to use these alternative habitats.

Warheit, Lindberg, and Boekelheide (1984) recognized three factors that determined the vulnerability of bird colonies to pinniped disturbance and destruction: 1) topographic relief, 2) pinniped densities, and 3) geographic location. The interference competition that we envision will only occur when both groups are simultaneously present and space is limited. If alternate breeding habitats are available to the birds (either physically or behaviorally), competition with pinnipeds would be reduced or nonexistent.

In this paper, we examine patterns that may represent different results or different stages of interplay among these three factors, both in the present and in the past. In the Recent we examine the intensity of the interaction at different geographic locations as topographies vary and the densities of pinnipeds differ. In the fossil record we look for patterns in seabird evolution, including diversity and geographic range, that coincide with the evolution and radiation of gregarious pinnipeds during the late Tertiary.

■

Recent Communities

Northern Hemisphere

The coast of central California supports almost half the breeding seabirds in the state (Sowls et al. 1980); the Farallon Islands, off San Francisco, alone support 35 percent of these birds. These islands also provide haulout habitat and breeding rookeries for three gregarious pinnipeds *(Zalophus californianus, Eumetopias jubatus,* and *Mirounga angustirostris)*. Año Nuevo Island, about 50 miles south of San Francisco, supports smaller populations of breeding seabirds (Sowls et al. 1980), but has larger pinniped populations relative to the island's size (LeBoeuf and Bonnell 1980).

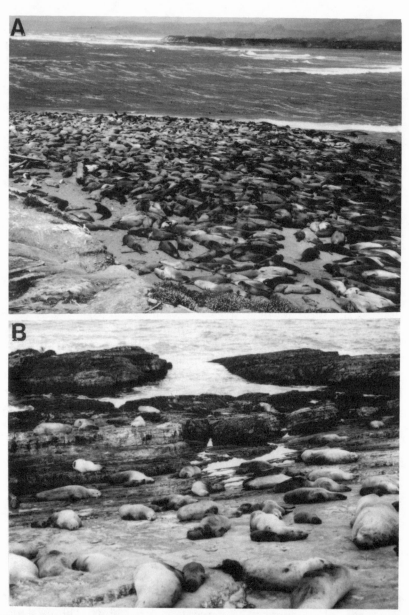

Figure 9.1. There are a number of ways in which gregarious pinnipeds occupy terrestrial space, depending upon species and time of year (i.e., whether they are breeding or molting). A) Postbreeding aggregation of northern elephant seals *(Mirounga angustirostris)* on sandy/cobble beach, Año Nuevo Island, California. B) Breeding female Steller's sea lions *(Eumetopias jubatus)*, also on Año Nuevo Island, are shown evenly spaced with their pups within males' territories. Photos by K. I. Warheit.

The pinniped populations on Año Nuevo and the Farallons have substantially increased during the past two decades (LeBoeuf and Bonnell 1980), and as a result, breeding shore- and seabirds have been disturbed by the pinnipeds' expansion into bird habitat. On Año Nuevo, oystercatcher chicks have been crushed and adult breeding activities and nest site defenses have been disrupted by gregarious pinnipeds. These factors have contributed to the decreased reproductive success of the birds (Warheit, Lindberg, and Boekelheide 1984). Brandt's cormorants *(Phalacrocorax penicilatus)* do not nest on Año Nuevo, but do roost in areas that appear to be adequate breeding habitats. At these roosting sites they are periodically scattered by subadult male Steller's sea lions *(Eumetopias jubatus)* that use these same areas. On Southeast Farallon, oystercatcher nest sites have also been destroyed by gregarious pinnipeds with increasing regularity (Warheit, Lindberg, and Boekelheide 1984), and on one broad marine terrace, Brandt's cormorants and common murres *(Uria alge)* have been displaced by California sea lions *(Zalophus californianus)* (Ainley and Boekelheide in press).

Although certain species of pinniped (especially the otariids) are nimble climbers, large breeding and haulout concentrations occur on flat areas close to shore (King 1983a). This means that seabirds that nest on flat areas near the shore are more vulnerable to pinniped disturbance than are birds that nest on steep slopes, cliffs, or high flat areas inaccessible to pinnipeds.

The interactions between birds and pinnipeds on Año Nuevo have been more intense than on Southeast Farallon because of the islands' respective topographies. Año Nuevo is a small, low, flat island 18 meters above sea level (Power 1980). In contrast, Southeast Farallon is a larger, precipitous (109 m high) granite peak, with a more heterogeneous landscape (Power 1980; DeSante and Ainley 1980). The only flat areas on the island are the marine terraces. On Año Nuevo, pinnipeds have access to the entire island; there is no space into which displaced breeding birds can move, except in the crevices formed by the different bedding planes of the shale. On Southeast Farallon the diverse topography provides pinniped-free refugia, and birds nesting in these areas may not be affected directly by increases in pinniped densities. The composition of seabird communities on these two islands may reflect these differences. On Southeast Farallon there are twelve breeding shore- and seabirds (Sowls et al. 1980), of which four (Brandt's cormorants, black oystercatchers, western gulls *[Larus occidentalis]*, and common murres) nest in areas that are potentially vulnerable to pinniped disturbance. On Año Nuevo there are five breeding shore- and seabirds (Sowls et al. 1980), and only the western gull and black oyster-catcher nests are vulnerable to pinnipeds.

Topography at or near breeding sites can be an important determinant of the relative intensity of interference competition between seabirds and gregarious pinnipeds. If pinnipeds encroach on a bird rookery, breeding birds may reduce or eliminate the interaction by moving to an area inaccessible to pinnipeds. This implies that alternative habitats are present and that birds have the behavioral flexibility to alter their breeding habitat preference. Because of tectonic activity and eustatic sea level change, the topography at seabird breeding localities is not static but varies throughout geologic time. When local sea levels are relatively high, breeding habitats may be reduced or eliminated, and interference competition between seabirds and pinnipeds for the limited space may result. Therefore, on a geologic time scale islands are created and destroyed repeatedly, and as such, both seabirds and pinnipeds are faced with inevitable habitat deterioration of varying intensities and durations.

Southern Hemisphere

Pinniped densities and geographic location are two other factors that contribute to the seasonal co-occurrence of pinnipeds and seabirds at particular island localities (Warheit, Lindberg, and Boekelheide 1984). Because breeding populations of both polygynous pinnipeds and seabirds are annually reestablished, moving the time of seabird arrival, breeding, and departure away from the time of pinniped activities should substantially reduce the competition. The degree to which these breeding features can be altered is constrained by other factors such as seasonal changes in food availability. Penguins are an excellent group to examine for temporal segregation of breeding.

Flightless birds like penguins are a special case in our interaction model, because flightless seabirds and pinnipeds both have limited terrestrial mobility. Flightless birds do not have the option of alighting from the air on inaccessible ledges. Like pinnipeds, they must amble overland between the ocean and their breeding sites.

Typically, large penguins are associated with broad, open breeding habitats. As penguin size decreases, breeding habitats become more variable to the point where steep slopes and cliffs become accessible habitats (table 9.1) (Stonehouse 1967); the smallest pen-

guins tend to be fossorial. Because pinnipeds typically occupy flat-
ter, more open habitats for breeding and hauling out (King 1983a),
we would predict that medium to large penguins would be the most
vulnerable to pinniped disturbance. Both gregarious pinnipeds and
penguins occur on the numerous islands north of and within the
Antarctic Convergence, including the "southern New Zealand islands" (Solander, Stewart, Snares, Auckland, and Campbell islands)
and South Georgia Island in the southern Atlantic.

Table 9.1 shows little or no overlap in breeding time or habitat
between pinnipeds and penguins on the southern New Zealand
islands; penguins tend to breed earlier in the year and in areas
inaccessible to large aggregations of pinnipeds, such as steep slopes
or dense forests. Only the rockhopper penguin *(Eudyptes crestatus)*
and the little penguin *(Eudyptula minor)* co-occur with and breed
at the same time as the gregarious pinnipeds *Phocarctos hookeri*
and *Arctocephalus forsteri.* These penguins breed on the more hilly
and rocky terrain or in burrows, whereas the pinnipeds are restricted to beach habitats. Although *P. hookeri* occasionally may be
found in dense vegetation up to half a mile inland (Crawley and
Cameron 1972), the largest concentrations occur on beaches (Falla,
Taylor, and Black 1979). Similar factors have also operated in South
Africa. Here, Rand (1951) has concluded that fur seal disturbance to
jackass penguins nesting on Seal Island, False Bay, would have been
more extensive if the penguins had not nearly completed breeding
before the seasonal increase in fur seal density. Rand (1951:102)
states that the "general effect [of fur seal disturbance] was to intro-
duce nesting uniformity in the [penguin] colony by prohibiting random nesting during the November/December [fur seal breeding sea-
son]."

As in New Zealand, the major breeding habitats of the penguins
on South Georgia Island are shifted to steeper and rougher terrain,
but there also appears to be considerable overlap on cobble beaches,
particularly between the king penguin *(Aptenodytes patagonia)* and
the fur seal *Arctocephalus gazella* (table 9.1). Because king penguins
breed in areas unoccupied by 95 percent of the South Georgia fur
seals (Croxall and Prince 1979; Hunter, Croxall, and Prince 1982),
this overlap may be exaggerated. Furthermore, Hunter, Croxall, and
Prince (1982:figure 9) show almost complete spatial segregation
between fur seal breeding areas and gentoo *(Pygoscelis papua)* and
macaroni *(Eudyptes chrysolophus)* penguin colonies on Bird Island,
South Georgia.

Rand (1954) reported that elephant seals *(Mirounga leonia)* may

Table 9.1. Height and breeding habits of Recent penguins on the Southern New Zealand Islands and South Georgia Island.

Species	Average Height	Breeding Time (JUL AUG SEP OCT NOV DEC JAN FEB MAR APR)	Breeding Habitat — TB	F	TS	BS	BC	BR	SR	SS	References[a]
New Zealand[b]											
Phocarctos hookeri		——— (NOV–JAN)				X					b,e,f,i
Arctocephalus forsteri		——— (NOV–DEC)						X			c,i
Megadyptes antipodes	66	—— (AUG–SEP)		X	X						a,h
Eudyptes robustus	56	—— (SEP)		X	X						a,h
E. pachyrhynchus	55	— (JUL)		X	X						a,h
E. crestatus	55		X						X	X	a,h
Eudytula minor	40	———— (JUL–DEC)	X	X							a,h
Mean =	54.40										
South Georgia Island											
Mirounga leonia		— (SEP)				X	X				i
Arctocephalus gazella		— (DEC)					X	X			d,g
Aptenodytes patagonia	95	———— (NOV–MAR)		X			X				a,h,j
Pygoscelis papua	81	—— (NOV)		X			X		X		a,h,j
Eudyptes chrysolophus	55	— (OCT)		X					X	X	a,h,j
Mean =	77.00										

NOTES: The table shows almost complete separation in breeding time and habitat between pinnipeds and penguins on the New Zealand Islands. Overall, smaller penguins appear to be either more flexible in their breeding habits than larger penguins, or occur in habitats not frequented by pinnipeds.

Abbreviations: TB = tunnels, burrows, or under vegetation; F = forest; TS = tussock, shrub; BS = beach sand; BC = beach cobbles/boulders; BR = beach, rocky; SR = slope, rolling; SS = slope, steep, or cliffs.

[a] a = Stonehouse 1967; b = Crawley and Cameron 1972; c = Crawley and Wilson 1976; d = Payne 1977; e = Falla, Taylor, and Black 1979; f = Wilson 1979; g = McCann 1980; h = Harrison 1983; i = King 1983a; j = Croxall and Prince 1980.

[b] Solander, Stewart, Snares, Auckland, and Campbell islands.

reduce the amount of potential breeding habitat for king penguins on Marion Island. However, on South Georgia deleterious interactions are avoided by segregation in time rather than in space; king penguin breeding commences after the elephant seals have finished their breeding and before their return to land for molt. Rand (1954) also suggested the king penguins could avoid interactions with the elephant seals by moving their incubated eggs. King penguins incubate their eggs on their feet, and so create somewhat of a portable nest site. This may enable them to share beach habitats with pinnipeds.

■
The Fossil Record

In this section, we examine two sets of fossil data for patterns that suggest that gregarious pinnipeds have been important components in the structuring of seabird evolution. The first set comes from the fossil record of pinnipeds and seabirds in the California Current upwelling system in the northeast Pacific. These data suggest that competition for space may have played a part in the extinction of certain seabirds, especially on offshore islands during Pleistocene high sea level stands. These extinctions subsequently shaped the composition of modern California seabird communities. The second data set suggests that the relatively recent arrival of gregarious pinnipeds into the ancient penguin systems of the Southern Hemisphere may have placed selective pressure on penguin size, a factor that can ultimately influence breeding habitat options in order to reduce competition for breeding space, as discussed above.

Before our fossil data sets can be used in such an analysis, the relative temporal occurrence of each seabird and pinniped species must be established. Fossils occur in rock units called formations. Biostratigraphic methods determine the relative temporal occurrence of each formation. Several biostratigraphic time scales are in use, and each is derived from the fossil record of the plankton (e.g., Foraminifera, radiolarians, diatoms), marine invertebrates, or land mammals. These time scales are subdivided into stages of zones, again by the presence or absence of key indicator species, and fossil-bearing formations are accordingly correlated with each other within a particular stage or zone. These stages are scaled to absolute time using radiometric dates when available. Using the biostratigraphic time scale and its associated radiometric dates, we are then able to plot the relative diversity of seabirds and pinnipeds on a common time scale. Furthermore, because the marine paleoclimatic and sea level data are established from the plankton or marine invertebrate data, we can overlay these data on the same time axis as the seabird and pinniped diversity curves. The resulting curves show, in geologic time, which pinniped and seabird species were contemporaneous, and what the major climatic regimes were during that time. The principal time scales used in this paper are from Berggren et al. (1985), Repenning and Tedford (1977), and van Eysinga (1978).

The fossil seabird data discussed below have been derived from the literature and from the work of Warheit (unpublished data). The fossil pinniped data are also derived from the literature and from L. G. Barnes (personal communication, 1986). The data for Tertiary climates, sea level, and geography are from the geological, micropaleontological and molluscan literature. Because the paleontological and geological literature for these topics is vast, only selected references appear in the figure and table captions. In this section the term "seabird" is used to include the Sphenisciformes, Procellariiformes, Pelecaniformes, Laridae, Alcidae, Gaviiformes, and Podicipediformes.

The California Tertiary

We grouped California fossil and Recent seabirds and pinnipeds into seven faunas, representing seven distinct units of time. The faunas range from about 23 million years ago (mya) to the Recent, with a hiatus from 21 mya to 16 mya (figure 9.2); each fauna corresponds approximately to a North American land mammal age. The temporal occurrence of each fauna ranges from about 2 million years (my) to 4.5 my. The difference in duration among faunas is, in part, an artifact of the available fossil record, as well as the precision of the correlations: rock units and index fossils generally provide better resolution in the Pliocene than in the Miocene. However, the longevity of each fauna may also represent long-term environmental stability; for example during the Miocene, the climate and sea level were more stable than in the Pliocene and Pleistocene, thereby producing more stable, longer-lived faunas. Like any fauna at any time, these paleofaunas are distinct entities subjected to specific biological and environmental conditions. Therefore, we consider them real, nonarbitrary ecological units.

The seabird and pinniped taxa occurring in each fauna are respectively summarized in tables 9.2 and 9.3 and in figure 9.3. The taxa listed for fauna R (Recent) are species that commonly occur in the California Current today; we did not restrict the analysis to breeding species only, because the breeding status of the fossil species could not be determined. However, by analogy to Recent Northern Hemisphere seabird communities, we assume that the Procellariiformes (Oceanitidae, and possibly the Procellariidae), Pelecaniformes, Laridae, and Alcidae represented in faunas O to V were breeding, and that the Gaviiformes, Podicipediformes, and some of the Procellariidae were only winter visitors. All pinniped species, with the possible exception of *Enaliarctos* and *Pinnartidion*, are assumed to be polygynous and gregarious (Mitchell 1965; Mitchell and Tedford 1973; Repenning 1976; see Barnes 1979 for sexual dimorphism in *Enaliarctos*).

The opposing diversity curves for pinnipeds and seabirds in figure 9.3 suggest that negative interactions between gregarious pinnipeds and seabirds may have affected seabird diversity along the California coast from 16 mya to the present. One alternative hypothesis is that both groups were responding oppositely to the same environmental variables, but as yet we can think of no variable that would be expected to produce opposite reactions. If the observed pattern resulted from competitive interactions, as we suggest, either food or space would have had to be limiting at certain times. Evidence

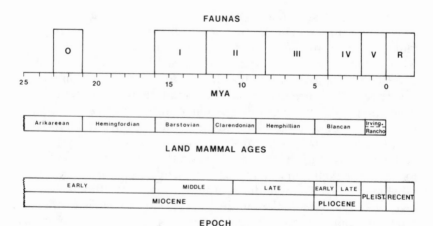

Figure 9.2. Correlation of California seabird/pinniped faunas with absolute time (in million years), land mammal ages, and geologic epochs. Correlations based primarily on Berggren et al. 1985, and Repenning and Tedford 1977.

for limiting food or space would support the seabird-pinniped competition hypothesis. We will examine these potential limiting factors, starting with space.

Increases in the number of islands and in the topographic complexity of these islands will increase available breeding habitats, and reduce potential competition for space. Likewise, a decrease in the number of islands will reduce breeding space, and increase the probability of competition. Tectonic processes and changes in sea level determine the topography of the nearshore terrestrial environment and especially islands. The relative changes in global sea level from the early Miocene to the Recent (Vail and Hardenbol 1979) are presented in figure 9.4. This curve shows a decline in sea level from a high stand during fauna I to a low stand in fauna III. This regression (low sea level) was followed by a major transgression (high sea

Table 9.2. Number of species in California Current seabird families through time.

	Fauna[a]						
	O	I	II	III	IV	V	R
Procellariiformes							
Diomedeidae	0	2	3	0	1	3	1
Procellariidae	0	4	5	2	2	3	6
Oceanitidae	0	0	0	1	1	0	5
Pelecaniformes							
Pelagornithidae	0	0	1	0	0	0	0
Pelecanidae	0	0	0	0	0	0	1
Sulidae	0	2	6	2	3	1	0
Phalacrocoracidae	0	0	1	0	2	2	3
Plotopteridae	1	0	0	0	0	0	0
Charadriiformes							
Alcidae	0	0	9	7	8	5	9
Laridae	0	0	0	0	3	2	18
Gaviiformes	0	0	1	1	1	3	3
Podicipediformes	0	0	0	0	4	5	4
TOTAL	1	8	26	13	25	24	50

SOURCES: Miller 1925, 1961; Wetmore 1930; Howard 1958, 1969, 1970, 1982, 1983 and references therein, 1984.
[a] Faunas from figure 9.2.

level) and then by increasingly rapid changes from fauna IV through fauna V. The Pleistocene (fauna V) shows the most frequent changes in sea level (figure 9.4); these are associated with the glaciation cycles. Vedder and Howell (1980) reconstructed middle to late Miocene California with numerous island archipelagos, and Cole and Armentrout (1979) described a Miocene and early Pliocene California coastline with numerous embayments. The reduction in island size and the loss of the island archipelagos and coastal embayments resulted from sea level transgressions, tectonic activity, and the Coast Range orogeny (Howard 1951). These changes drastically reduced the nearshore terrestrial environment from the Pliocene to the Recent (late fauna III and fauna V) (table 9.4). We envision space to be most limiting and competition most intense in these faunas. We see major declines in seabird diversity during these times (figure 9.3).

To demonstrate that trophic competition was a determinant in

Table 9.3. Number of species of gregarious pinnipeds in the California Current through time.

	Fauna[a]						
	O	I	II	III	IV	V	R
Desmatophocids							
Allodesminae	0	4	1	0	0	0	0
Desmatophocinae	0	2	1	0	0	0	0
Pinnartidion	1	0	0	0	0	0	0
Odobenids							
Odobeninae	0	0	0	2	0	0	0
Imagotariinae	0	1	1	2	0	0	0
Otariids							
Otariinae	0	0	1	3	3	3	3
Enaliarctos	2	0	0	0	0	0	0
Valenictus	0	0	0	1	0	0	0
Phocids	0	0	0	0	0	1	1
Mirounga	0	0	0	0	0	1	1
TOTAL	3	7	4	8	3	4	4

SOURCES: Barnes 1972, 1979; Mitchell and Tedford 1973; Barnes and Mitchell 1975; Repenning and Tedford 1977; Barnes, Domning, and Ray 1985.
NOTES: Taxa listed follow, with a few exceptions, the classification of Barnes 1979.
[a] Faunas from figure 9.2.

the evolution of the California seabird communities, it must be shown that food was periodically limited in geologic time. Ocean temperatures in the California Current have changed in the past 25 million years, cooling from a subtropical environment in the early Miocene to a cold-water upwelling system today (Addicott 1970; Keller and Barron 1981; Mullineaux and Westberg-Smith 1986). Keller and Barron (1981), Mullineaux and Westberg-Smith (1986), and others have estimated that the onset of Antarctic glaciation at about 15 mya (fauna I) initiated the cold-water upwelling in the California Current. This transformation from a warm-water to cold-water upwelling environment should have created a more concentrated and predictable food source and therefore enhanced the survival of seabirds that typically use upwelling systems. In particular, diving birds (i.e., alcids, cormorants, loons, and grebes) and some sulids (e.g., species similar to the Peruvian booby [*Sula variegata*] and the cape gannet) should have benefited most from this change (see Ainley [1977] for seabird feeding habitats). This is consistent with the cormorant, loon, and the grebe data in Table 9.2.

Figure 9.5 contrasts the diversity patterns among pursuit plungers/surface seizers (Procellariiformes), plungers (Sulidae), and diving seabirds. Divers show a general increase in diversity through time,

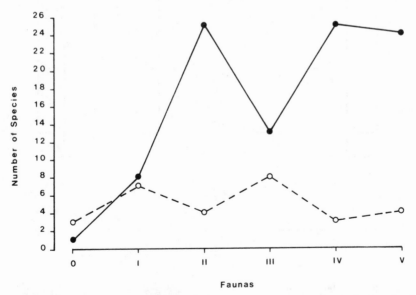

Figure 9.3. Number of species of seabirds (solid line) and pinnipeds (broken line) from fauna 0 to fauna V. Data from tables 9.2 (seabirds) and 9.3 (pinnipeds).

while the sulids show a decline in diversity from fauna II (with six species) to local extinction today. Although the pursuit plungers are more difficult to interpret, they show a significant increase in diversity ($b = 1.179$; $t = 3.684$; $p < .01$; $r = .682$; $df = 5$). The general decline by all foraging groups during fauna III may be the result of trophic competition between seabirds and marine mammals. However, the paleoclimatic data summarized above show no decline in upwelling during fauna III. In fact, Keller and Barron (1981) describe

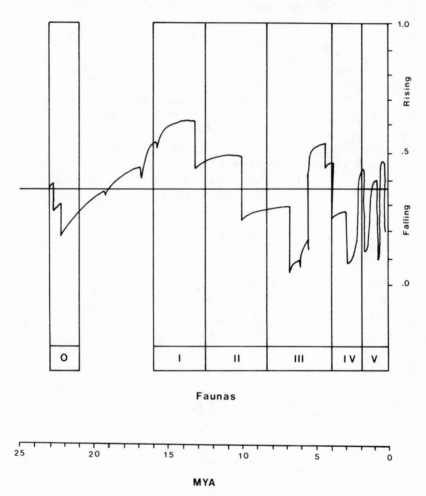

Faunas

MYA

Figure 9.4. Cycles of global sea level change scaled to absolute time (in million years) and California seabird/pinniped faunas. Present-day sea level is represented by horizontal line at center of graph. Sea level data from Vail and Hardenbol 1979.

two cold climate events (with pulses of intense upwelling) during fauna III, which suggest that food was not limited. Furthermore, if we assume that trophic structure was the main determinant of the diversity curves shown in figures 9.3 and 9.5, we would expect that all diving birds (at least all the piscivorous diving birds) would show similar patterns. This is not the case. When diving birds are divided

Table 9.4. Estimated length of California coastline from Miocene to Recent.

Epoch	Length (km)
Miocene	1,714
Pliocene	2,057
Pleistocene	1,490
Recent	1,286

SOURCES: Data for Miocene and Pliocene from figures 3 and 4 of Cole and Armentrout 1979; data for Pleistocene and Recent from Economic Mineral Map of California, no. 7, Lead and Zinc, *Calif. J. Mines Geol.* (1957), vol. 53, nos. 3–4.

NOTE: Pleistocene coastline equal to 100-meter contour.

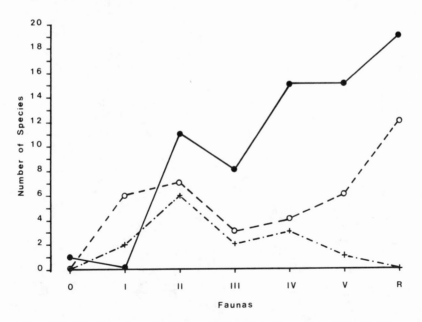

Figure 9.5. Number of species of pursuit plungers/surface seizers (Procellariiformes) [broken line], plungers (sulids) [dotted broken line], and divers [solid line] from fauna 0 to fauna R. Foraging classes follow Ainley 1977. Data from table 9.2.

into flighted and flightless groups we see different diversity patterns (figure 9.6). While the volant divers' diversity curve closely resembles the total seabird curve (figure 9.3), the flightless divers show an almost symmetric diversity, with a peak during fauna III. However, there is an inverse relationship between the diversity of flightless divers and sea level, with diversity decreasing as sea levels rise (compare figure 9.6 with figure 9.4).

As we discussed above, flightless seabirds are a special case because they share many land-based characteristics with pinnipeds and are thus especially vulnerable to interference competition. Three groups of flightless birds have inhabited California waters since the Miocene: 1) the plotopterids, 2) the mancallines, and 3) the "scoter" *Chendytes*.

The plotopterids (table 9.2) were flightless Pelecaniformes closely related to the Sulae (Olson 1985b). The group became extinct in the early Miocene (fauna 0) during the initial pinniped radiation *(Enaliartos* and *Pinnartidion)* (table 9.3). Olson and Hasegawa (1979) have suggested that the plotopterids' extinction may have been precipitated by the evolution and radiation of pinnipeds and ceta-

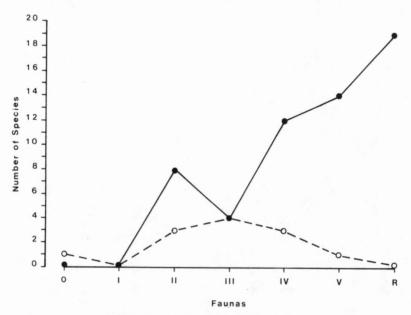

Figure 9.6. Number of species of volant (flighted) [solid line] and flightless [broken line] diving seabirds from fauna 0 to fauna R. Data from table 9.2. Data for flightless alcids (mancallines) from literature cited in table 9.2.

ceans. Although there is some question concerning the breeding habits of the enaliarctids (Mitchell and Tedford 1973; Barnes 1979), the Hemingfordian hiatus in the California marine vertebrate fossil record (figure 9.2) may mask the contemporaneous occurrence of plotopterids and sexually dimorphic/gregarious pinnipeds in California; the plotopterids enter the hiatus and the large, sexually dimorphic Desmatophocidae emerge from it. Plotopterids also may have competed with desmostylids for nearshore terrestrial space. Desmostylids were amphibious, hippopotamuslike marine mammals that occurred along the coastal areas of the northern Pacific (see Barnes, Domning, and Ray 1985 for review of fossil record). Because the large plotopterids of Japan may already have disappeared before the Hemingfordian hiatus (S. L. Olson, personal communication, 1986), their extinction would have to be attributed to some other phenomenon. As Olson and Hasegawa (1979) suggested, perhaps competition for food with cetaceans may have been important.

The shape of the diversity curve for flightless diving seabirds in California is mainly determined by a group of extinct alcids known collectively as the mancallines (genera: *Mancalla, Praemancalla,* and *Alcodes*). Their life histories and breeding habitats are thought to have been similar to those of the extant alcids, but because they were flightless, they undoubtedly shared characteristics with the flightless great auk *(Pinguinus impennis)* of the northern Atlantic. Great auks bred on low-lying islands and gently sloping shores where access was not limited or difficult (Bengstson 1984; Harris and Birkhead 1985). This same type of habitat would be favored by gregarious pinnipeds and the birds occupying it would be particularly vulnerable to spatial competition.

The mancallines evolved and radiated during the periods of decreasing sea levels of faunas II and III, and began their decline during the second half of fauna III (figures 9.4, 9.6). During fauna IV the group was still abundant (but less diverse) in the San Diego area, with almost 2,000 specimens recovered from the San Diego Formation (Chandler 1985, personal communication, 1986). The group underwent a precipitous decline in fauna V, with only a single specimen found from a Pleistocene deposit dated at approximately 470,000 years ago (Howard 1970, 1982; Kohl 1974). *Mancalla* may have also persisted into the Pleistocene of Japan (S. L. Olson, personal communication, 1986). The group may have finally gone extinct during the Pleistocene high-water Yarmouth interglacial period.

If the mancallines and the Great Auk both bred in habitats vulnerable to disturbance by gregarious pinnipeds, why did the great auks survive well into the Recent? The answer may be that there has never been any Otariidae or other similarly gregarious pinniped near great auk breeding colonies. The only published records of a fossil otariid from the Atlantic Ocean north of the equator are fragmentary remains from North Carolina and Virginia referred to the odobenid genus *Prorosmarus* (see Repenning and Tedford 1977). All other otariid fossil taxa are from either the Southern Hemisphere or the northern Pacific (Repenning and Tedford 1977; Repenning, Ray, and Grigorescu 1979; Barnes, Domning, and Ray 1985). The walrus *(Odobenus rosmarus)* occurs in the northern Atlantic today, but in areas far to the north of any of the known great auk colonies. Fossil phocid pinnipeds have occurred in the northern Atlantic (Ray 1976; Repenning, Ray, and Grigorescu 1979), but because they are not known to be sexually dimorphic, we conclude that none were gregarious (see Bartholomew 1970 for the association between sexual dimorphism and gregarious breeding in pinnipeds). Furthermore, except for the walrus mentioned above, and the winter breeding gray seal *(Halichoerus grypus)*, no land-based (versus ice-inhabiting) gregarious pinnipeds exist in the northern Atlantic today (King 1983a). Thus, great auks never interacted with gregarious otariids, and their survival into historic times lends support to our seabird-pinniped interaction hypothesis.

The last group of flightless marine birds in California were two species of Anseriformes in the genus *Chendytes*. The breeding habits of these species are not known and comparisons with modern flighted anatids may be tenuous. Perhaps their breeding habits were most similar to the flightless steamer ducks (genus: *Tachyeres*) of South America. These ducks are not gregarious, and nest under vegetation or among grass or dry kelp (see Johnsgard 1978, and references therein). *Chendytes* is known only from the late Pleistocene of southern California, and survived into historic times. Because their bones are often found in American Indian kitchen middens, their extinction has been associated with human predation (Morejohn 1976).

In summary, diversity curves of seabirds and pinnipeds in the California Miocene to Recent show contrasting patterns. Declines in seabird diversity occur during times of increasing pinniped diversity and increasing terrestrial habitat deterioration (e.g., rising sea level). This occurred during faunas III and V, and is consistent with the pattern predicted if space was a critical resource for both pinnipeds and seabirds in the nearshore marine environment. When seabirds are divided into foraging classes, the volant divers and pursuit plungers show increases in diversity through time. All foraging classes show a decrease in diversity during fauna III, but only the plungers (sulids) and flightless divers show declines during fauna V. The dramatic rise in sea level midway through fauna III, and the rapid changes in sea level in fauna V, combined with increased pinniped diversity, may have created these patterns. Thus, declines in seabird diversity are concentrated in two groups, flightless seabirds and sulids. Whereas flightless seabirds have a unique vulnerability to pinniped disturbance, the flighted seabirds such as sulids do not, and the regional extinction of the sulids appears to be a more complicated issue (Warheit, unpublished data). Sulid extinction in the northeast Pacific may have involved a combination of factors, including an inability to behaviorally respond to the dramatic and rapid changes in their breeding habitats during the Pleistocene glaciation; pinniped disturbance; and degree of philopatry (Warheit, unpublished data).

Tertiary Penguins and Pinnipeds

If deleterious interactions between flightless seabirds and gregarious pinnipeds in the Tertiary of California lead, in part, to certain seabird extinctions, the arrival of gregarious pinnipeds in the Southern Hemisphere should also have produced changes in the penguin faunas. Penguin diversity from the middle Eocene to the present is shown in figure 9.7. Like the time scale for the California system discussed above, absolute time is divided into four faunas, with fauna 1 the oldest and fauna 4 being that of the Recent. The penguin fossil record is extensive, incorporating over 40 million years and five geographical provinces. The earliest arrival of any type of pinniped into the Southern Hemisphere was much later (middle Miocene) than the Eocene appearance of penguins (table 9.5). Moreover, the first co-occurrence of pinnipeds with penguins is even later, sometime during the late Miocene/early Pliocene of South Africa and Peru (figure 9.7, table 9.5).

Because penguins are a food resource for many species of pinnipeds today (for example, see Bonner and Hunter 1982), the arrival of a new predator in the system would undoubtedly alter penguin foraging behavior; but did the addition of major space occupiers also drastically change penguin breeding behavior? This question cannot be answered directly from the fossil record because it is impossible to determine the breeding biology of extinct penguins, and except for the genera *Aptenodytes* and *Pygoscelis*, all other fossil penguins are currently placed in extinct genera, making comparisons with recent taxa tenuous. However, as discussed above, there does appear to be an association between penguin breeding habitats and penguin size (height) (table 9.1). These data show that smaller birds use more diverse breeding habitats, or occupy areas inaccessible to pinnipeds. Because of the greater range in breeding habitats associated with

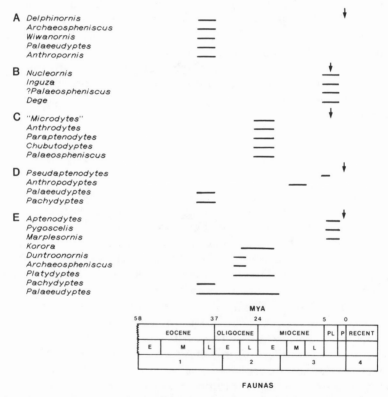

Figure 9.7. Geographic and temporal distribution of fossil penguin genera. Time measured in million years, geologic epochs, and penguin faunas (1–4). Arrows show time of first arrival of pinnipeds for each geographic region. Group A for Seymour Island, Antarctica; B for South Africa; C for South America (Patagonia); D for Australia; and E for New Zealand. Time correlation based on van Eysinga 1978, and Berggren et al. 1985. Penguin data from Simpson 1965, 1970, 1971, 1975, 1981; Jenkins 1974; Tonni 1980; Olson 1985a; and references therein. Pinniped data from table 9.5. The genus "Microdytes" (c) has been changed to *Eretiscus* (Olson 1986).

small size, we considered smaller penguins to be better equipped to deal with pinniped disturbance than larger penguins. By selecting for increased breeding habitat flexibility, pinniped disturbance may be one of several factors that influences penguin size. Therefore, changes in penguin breeding habitats in response to the arrival of pinnipeds may be suggested by examining penguin size through time.

Mean heights of fossil penguins were calculated and are presented in table 9.6. Only the New Zealand/Australia group provides a complete series from fauna 1 to fauna 4. Mean penguin height in fauna 1 is larger than mean heights in all other faunas. The mean heights for New Zealand/Australian penguins are consistently the largest for all geographic regions during all faunas except fauna 4 (the Recent), where they are the smallest. Finally the smallest mean height for penguins in New Zealand/Australia and the Antarctic occurs during the faunas in which pinnipeds first arrived (both fauna 4). Although there are no penguin fossils from fauna 3 of South America, there is no difference between penguin heights in fauna 2 and fauna 4. The data for South Africa are difficult to interpret because of small sample sizes in the two faunas.

The trend in mean height of New Zealand/Australian fossil penguins is particularly interesting in view of the data in table 9.1. The

Table 9.5. Fossil pinnipeds from the Southern Hemisphere.

Species	Period	Location	References
Phocids			
Properiptychus argentinus	Middle Miocene	Argentina	b
Homiphoca capensis	Late Miocene/ early Pliocene	South Africa	a,c
Acrophoca longirostris	Early Pliocene	Peru	d
Piscophoca pacifica	Early Pliocene	Peru	d
Otariids			
Arctocephalus lomasiensis	Early Pliocene	Peru	e
Neophoca palatina	Middle Pleistocene	New Zealand	f

SOURCES: a: Hendey and Repenning 1972; b: de Muizon and Bond 1982; c: de Muizon and Hendey 1980; d: de Muizon 1981; e: de Muizon 1978; f: King 1983b.

NOTE: All other fossil pinnipeds from the Southern Hemisphere are of Recent species and no older than late Pleistocene.

Table 9.6. Penguin height through time.

GEOGRAPHIC REGION	Fauna			
	1	2	3	4
New Zealand/Australia	140.50 (9.18) [4] [a]	90.50 (20.32) [8]	93.50 (14.27) [4]	54.40 (9.29) [5] [b]
South America	—	76.20 (24.52) [10]	—[b]	74.20 (14.86) [5]
South Africa	—	—	58.75 (10.97) [4] [b]	70 [1]
Antarctic Peninsula	111.83 (24.07) [6]	—	—	80.80 (19.79) [5] [b]
MEAN HEIGHT	123.30 (23.87) [10]	82.56 (23.27) [18]	76.13 (22.00) [8]	70.80 (17.83) [15]

SOURCE: From literature cited in figure 9.7.

SOURCES: Simpson 1965, 1970, 1971, 1975, 1981; Jenkins 1974; Tonni 1980. Olson 1985a, and references therein. Pinniped data from table 9.5.

[a] Mean height (standard deviation) [n] of penguins in geographic region for each fauna.

[b] Denotes first arrival of pinnipeds into region.

pattern (table 9.5, figure 9.8) is consistent with the idea that the appearance of pinnipeds would provide a selective force for reducing penguin height and increasing penguin breeding habitat breadth. The decrease in mean penguin height from fauna 3 to fauna 4 was also accompanied by an increase in sea surface temperature (figure 9.8). This is consistent with the observation that larger penguins occur in colder waters, and smaller penguins occur in warmer waters. Therefore, changes in water temperature may also be responsible for the temporal patterns seen in small and large penguins. However, sea surface temperatures during faunas 1 and 2 are considerably warmer than that during fauna 4, yet their penguin sizes are significantly larger than mean penguin size in fauna 4. These results run counter to those predicted from the thermal model. Simpson (1976) also recognized this problem and further emphasized that the large penguins of the Eocene and Oligocene (fauna 1) lived in seas considerably warmer than those inhabited by the largest penguins today (the emperor penguin [*Aptenodytes forsteri*] and king penguin).

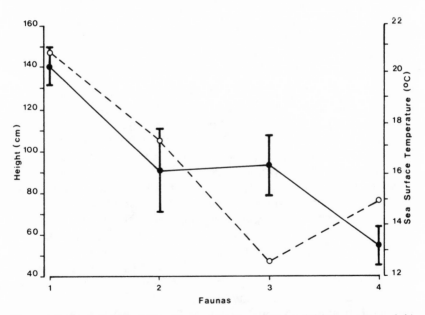

Figure 9.8. Mean height of New Zealand fossil penguins (solid line; y-axis on left) and mean sea surface temperatures (broken line; y-axis on right) for faunas 1 through 4. Penguin mean heights shown with 1 standard deviation. Penguin data from literature cited in figure 9.7. Sea surface temperature data from Devereux 1967.

The data for South America are not consistent with the pinniped disturbance model. However, as stressed above, interference competition only occurs when there is a shortage of space. Because the availability of breeding space in geologic time is a product of both local (e.g., tectonic subsidence) and global (eustatic sea level changes) events, it is possible that space has never been a critical problem in South America and therefore pinnipeds have never been an important selective force of structuring mechanism in South American penguin communities.

In summary, the history and distribution of flightless seabirds through time have been affected by the following pinniped related, large-scale biogeographic patterns: 1) the evolution and radiation of pinnipeds in the northern Pacific Ocean and, during times of habitat deterioration, extinctions of flightless seabirds (plotopterids and mancallines); 2) the arrival of gregarious pinnipeds in the Southern Hemisphere and height reductions in flightless seabirds (penguins); and 3) in the absence of gregarious pinnipeds, the evolution and survival of a flightless seabird in the northern Atlantic Ocean (great auks).

■

Discussion

We have used both Recent and fossil data to form and test hypotheses about the importance of seabird-pinniped spatial competition in structuring seabird communities. Biogeographic patterns and species interactions from Recent island communities form the basis of our hypotheses. These hypotheses yield predictions about the history and evolution of these seabird communities, and the predictions are compared directly with the fossil record. The fossil record, therefore, provides the test for evolutionary hypotheses derived from Recent ecological data. Pregill and Olson (1981:93) argue correctly that "fossils . . . provide concrete evidence that bears directly on ecological theory in showing the importance of historical events in shaping the distributions and adaptations of organisms." We extend this claim to populations and communities, and show that the historical events may be biological (e.g., the evolution and dispersal of gregarious pinnipeds) and/or physical (e.g., changes in relative sea levels). It is our opinion that the need to observe and document processes in the context of an *evolutionary* setting, which requires a temporal view beyond ecological time, cannot be over-emphasized when workers seek to understand how the past has shaped the present. Many of the habitats and species that we observe, manipulate, and document today have been substantially altered by interactions with humans since the Pleistocene (see Olson and James 1982a, 1982b; Steadman and Olson 1985). This interaction can mask or eliminate the selective forces that were once prominent for eons of time as well as produce new selective forces that have not yet shaped modern communities. Furthermore, climate has also drastically changed since the Pleistocene, and many biogeographic patterns may reflect climates of the Pleistocene rather than of the Recent epoch. For example, Pregill and Olson (1981) showed that climates and sea levels during the Pleistocene explain modern West Indian vertebrate species distributions and endemicity better than do the theoretical taxon cycles proposed by Ricklefs and Cox (1972).

For the greater part of this century, seabird and pinniped densities have been depressed worldwide to the point at which space has not been a critical factor in structuring communities. This situation is an artifact of human exploitation (e.g., the harvesting of seals and guano, and the collecting of eggs), and does not represent the environmental conditions in which present-day seabird species and their communities evolved. Prior to human involvement in the nearshore marine environment, space may have been an important criteria in structuring communities. The fossil record suggests that during periods of rapid environmental change the diversity of certain groups of seabirds has also changed. Not all groups were affected equally, primarily because different groups use space differently. Furthermore, some species were buffered from space shortages and interference competition (pinniped or even seabird interspecific) by use of alternative habitats. If this is true, then limiting breeding space may strongly select for those species with greater flexibility in breeding requirements. Climatic changes operating in ecological time (e.g., El Niño events) may be partly responsible for changes in local seabird densities, but we feel that the climatic events that operate over geologic time are the forces that produce the relative species compositions of seabird communities. For example, the increasingly rapid fluctuations in sea levels during the Pleistocene, which produced rapidly changing coastal environments, may have been a selective environment in which only habitat generalists survived. With the compounding effects of pinniped disturbance on the low flat habitats, only those seabird species capable of nesting on steep slopes of cliffs were able to breed suc-

cessfully. Seabird species composition at different breeding localities today may reflect this process. For example, Warheit (unpublished data) has observed that seabirds currently breeding on Año
Nuevo Island either nest on the small, low ledges or crevices scattered along the perimeter of the island (pigeon guillemots [Cepphus
columba] and pelagic cormorants [Phalacrocorax pelagicus]), or are
habitat generalists (western gulls [Larus occidentalis]). As mentioned above, Brandt's cormorants only roost on the island, and at
these roosting sites they are periodically disturbed by Steller's sea
lions. (Cape cormorants [P. capensis] roost, but do not breed on Seal
Island, False Bay, South Africa, and are similarly affected by fur
seals [Rand 1951]). By inhabiting a low flat island with a high
pinniped density, the seabird community on Año Nuevo Island may
represent the seabird communities of the California Pleistocene.

On South Georgia, pinniped disturbance appears to be a selective
force favoring species with particular habitat preferences and behaviors. The dramatic increase in Antarctic fur seals, from a few hundred
in the 1930s to over 800,000 today, has resulted in seabird breeding
habitat destruction (Hunger, Croxall, and Prince 1982; Croxall et al.
1984). Croxall et al. (1984:658) report that there are only a "few
areas of flat coastal ground at Bird Island with any breeding petrels,
and only the interior valleys and very steep slopes remain undis-
turbed." Diving petrels (genus: Pelecanoides) are unaffected by the
fur seal disturbance because they tend to breed on steeper terrain,
and skuas (genus: Catharacta), traditionally restricted to meadow
habitats, have moved into the disturbed tussock grasslands (Croxall
et al. 1984).

On Hollamsbird Island, Namibia, Rand (1952:456) states that
"subadult [fur seals] were occasionally seen 'threatening' the cormorants standing on the cliff edge, but as they could not climb over
this precipitous area they did not seriously affect these birds." And
on Seal Island, False Bay, South Africa, Rand (1951) has observed
that the ground-nesting birds are most affected by pinniped disturbance. Of these ground-nesting species, those completing their nesting
prior to the arrival of seals (e.g., jackass penguins) are less affected
than those nesting during the fur seal breeding season (e.g., kelp
gulls [Larus dominicanus]). Rand concludes (1951:103) that "among
the gulls ... the effect of the [fur] seal herd is more serious, and
unless there is some change in their nesting habits these birds will
cease to breed successfully on the island."

We do not believe that all seabird extinctions that have simultaneously occurred with gregarious pinniped radiations result solely

from interference competition, or that spatial competition is the only important factor in structuring seabird communities. We suggest only that the deleterious interactions are real and have occurred in the past, and like a host of other selective factors, have *contributed* to the evolution of modern seabird communities.

If protection for marine mammals is maintained or increased and if marine birds become protected in a similar way, population densities could continue to rise at fast rates. And with the continuing increase of urbanization and recreational use of the coastal zone by people, the amount of available breeding and haulout space will also decline. Coupling the effects of increased protection for marine vertebrate populations and declining habitat should produce an increase in interactions between gregarious pinnipeds and seabirds at seabird breeding localities. These occurrences should not be considered isolated, interesting phenomena, but important structuring forces in past, and possibly future seabird communities. Moreover, they point out the need for careful consideration and modeling of the implications of wildlife management and protection for *all* species in the system, and not just those with which we share a more recent descent.

■
Summary

In this paper we have used patterns from fossil and Recent marine communities to test hypotheses on the importance of pinniped-seabird spatial competition in structuring seabird communities. Spatial interactions between seabirds and pinnipeds have been overlooked for at least two reasons. First, because Recent populations of seabirds and pinnipeds are so depressed as a result of human activities (indigenous and modern societies), the frequency of spatial interactions is relatively low. Second, large-scale competitive interactions influencing seabird community structure may only occur during times of environmental change measured on a geologic time scale, rather than an ecological time scale.

The interactions between pinnipeds and shore- and seabirds at various Recent nesting localities along the North American, South American, and African coasts observed by us or published in the literature alerted us to the potential importance of spatial interactions in structuring seabird communities. We observed that not all pinnipeds are potential spatial competitors with seabirds. Only land-based gregarious pinnipeds, namely the otariids and some phocids (e.g., elephant seals) form haulout or breeding aggregations large enough to have a significant impact on seabirds. Furthermore, these pinnipeds are sexually dimorphic, and are therefore recognizable in the fossil record. Likewise, not all seabirds are vulnerable to pinniped disturbance. Because gregarious pinnipeds occupy low, flat areas near the water's edge, only those shore- and seabirds restricted to this habitat will be affected.

From this we make a number of predictions. First, topographic relief is an important variable. If an island has little or no topographic relief, all areas may be accessible to pinnipeds. Spatial interactions will be more intense here than on islands with a heterogeneous landscape. Because island topography is a product of geologic factors, habitat deterioration is an inevitable fact of every seabird colony. Second, flightless seabirds (e.g., penguins) are most vulnerable to pinniped disturbance due to their limited terrestrial mobility. Third, seabirds can reduce the level of disturbance by moving to habitats inaccessible to pinnipeds—provided that this habitat is either physically or behaviorally available—or by changing their time of breeding to a period when pinniped densities are lower.

We examined two situations in Recent seabird communities. In California, gregarious pinnipeds have disrupted black osytercatchers, common murres, and Brandt's cormorants breeding on Año Neuvo and Southeast Farallon Island. Año Nuevo has been impacted more intensely than Southeast Farallon because the entire island is accessible to pinnipeds. The heterogeneous landscape of Southeast Farallon provides habitat refugia for nesting seabirds. In the Southern Hemisphere, there is almost complete temporal and/or spatial separation between penguins and pinnipeds breeding on some New Zealand islands, on South Georgia Island, in the southern Atlantic, and in South Africa.

The fossil record reveals similar patterns. In the California Current upwelling system, seabird diversity declined during periods of increased pinniped diversity and nearshore terrestrial habitat deterioration. All groups were not affected equally, with declines in flightless alcids (mancallines) and sulids being the most severe. In the Southern Hemisphere, penguins, especially those from New Zealand/Australian areas, showed substantial shifts to smaller body sizes from the late Eocene to the present. These shifts occurred during the time of first local appearance of gregarious pinnipeds. Because body size is inversely related to breeding habitat diversity in Recent penguins, a shift to smaller body sizes suggests that the arrival of pinnipeds was a factor selecting for species with greater breeding habitat diversity. This would decrease the frequency of deleterious interactions between the penguins and pinnipeds through time, and allow for coexistence today.

Finally, seabird communities are long-lived historical entities that have been influenced by past environments with different selective regimes. Modern-day species composition and structure in these communities are, in part, a product of these historical processes. Therefore, documenting seabird community structure as an evolutionary process requires a temporal view beyond ecological time.

Acknowledgments

We thank David Ainley, Lawrence Barnes, Joanna Burger, James Estes, Cheryl Niemi, Storrs Olson, and Kevin Padian for comments on the manuscript; Robert Chandler and Storrs Olson for sharing with us their unpublished fossil seabird data; and Lawrence Barnes for reviewing the fossil marine mammal data, and for providing us with Southern Hemisphere pinniped references. We also thank Leo Laporte for early discussions of our initial hypotheses, and Joanna Burger for inviting us to participate in this book and giving us the opportunity to express our ideas. Figures were drafted by Jack Blazek. Partial funding for this project was given to K. Warheit by the Remington Kellogg Fund of the Museum of Paleontology, University of California, Berkeley.

References

Addicott, W. O. 1970. Tertiary paleoclimatic trends in the San Joaquin Basin, California. *U.S. Geol. Surv. Prof. Paper* 644-D:D1–D19.

Ainley, D. G. 1977. Feeding methods of seabirds: a comparison of polar and tropical communities. In G. A. Llano, ed., *Adaptations in Antarctic Ecosystems*, pp. 669–685. Houston, Tex.: Gulf.

Ainley, D. G. and R. J. Boekelheide eds. In press. *The Breeding Ecology of Seabirds: Perspectives from the Farallones.* Stanford University Press.

Ainley D. G., R. E. LeResche, and W. J. L. Sladen. 1983. *Breeding Biology of the Adelie Penguin.* Berkeley: University of California Press.

Ainley, D. G. and T. J. Lewis. 1974. The history of Farallon Island marine bird populations, 1854–1972. *Condor* 76:432–446.

Barnes, L. G. 1972. Miocene Desmatophocinae (Mammalia: Carnivora) from California. *Univ. Calif. Publ. Geol. Sci.* 89:1–68.

Barnes, L. G. 1979. Fossil Enaliarctine pinnipeds (Mammalia: Otariidae) from Pyramid Hill, Kern County, California. *Contrib. Sci. Nat. Hist. Mus. Los Angeles Co.* 318:1–41.

Barnes, L. G., D. P. Domning, and C. E. Ray. 1985. Status of studies on fossil marine mammals. *Mar. Mam. Sci.* 1:15–53.

Barnes, L. G. and E. D. Mitchell. 1975. Late Cenozoic northeastern Pacific Phocidae. *Rap. P.-v. Reun. Cons. int. Explor. Mer.* 169:34–42.

Bartholomew, G. A. 1970. A model for the evolution of pinniped polygyny. *Evolution* 24:546–559.

Bengstson, S. 1984. Breeding ecology and extinction of the Great Auk *(Pinguinus impennis):* anecdotal evidence and conjectures. *Auk* 101:1–12.

Berggren, W. A., D. V. Kent, J. J. Flynn, and J. A. Van Couvering. 1985. Cenozoic geochronology. *Geol. Soc. Amer. Bull.* 96:1407–1418.

Bonner, W. N. and S. Hunter. 1982. Predatory interactions between Antarctic Fur Seals, Macaroni Penguins and Giant Petrels. *Br. Antarct. Surv. Bull.* 56:75–79.

Chandler, R. 1985. Fossil birds from the San Diego Formation, Late Pliocene, Blancan, San Diego County, California. Thesis, San Diego State University, California.

Cole, M. R. and J. M. Armentrout. 1979. Neogene paleogeography of the western United States. In J. M. Armentrout, M. R. Cole, and H. Terbest, Jr., eds., *Cenozoic Paleogeography of the Western United States* (Pacific Coast Paleogeography Symposium 3), pp. 297–323. Los Angeles: Society of Economic Paleontologists and Mineralogists.

Coulson, J. C. and C. S. Thomas. 1985. Changes in the biology of the Kittiwake *Rissa tridactyla:* a 31-year study of a breeding colony. *J. Anim. Ecol.* 54:9–26.

Crawley, M. C. and D. B. Cameron. 1972. New Zealand sea lions, *Phocarctos hookeri,* on the Snares Islands. *N. Z. J. Mar. Fresh. Res.* 6:127–132.

Crawley, M. C. and G. J. Wilson. 1976. The natural history and behavior of the New Zealand Fur Seal *(Arctocephalus forsteri). Tuatara* 22:1–29.

Croxall, J. P. and P. A. Prince. 1979. Antarctic seabird and seal monitoring studies. *Polar Record* 19:573–595.

Croxall, J. P. and P. A. Prince. 1980. Food, feeding ecology and ecological segregation of seabirds at South Georgia. *Biol. J. Linn. Soc.* 14:103–131.

Croxall, J. P., P. A. Prince, I. Hunter, S. J. McInnes, and P. G. Copestake. 1984. The seabirds of the Antarctic Peninsula, islands of the Scotia Sea, and Antarctic Continent between 80°W and 20°W: their status and conservation. In J. P. Croxall, P. G. H. Evans, and R. W. Schreiber, eds., *Status and Conservation of the World's Seabirds,* pp. 637–666. Cambridge: International Council for Bird Preservation.

DeSante, D. F. and D. G. Ainley. 1980. The avifauna of the South Farallon Islands, California. *Stud. Avian Biol.* 4:1–104.

Devereux, I. 1967. Oxygen isotope paleotemperature measurements of New Zealand Tertiary fossils. *N. Z. J. Sci.* 10:988–1011.

Diamond, A. W. 1979. Dynamic ecology of Aldabran seabird communities. *Phil. Trans. R. Soc. Lond. B.* 286:231–240.

Diamond, A. W. and C. J. Feare 1980. Past and present biogeography of central Seychelles birds. *Proc. 4 Pan-Afr. Orn. Congr.,* pp. 89–98.

Eysinga, F. W. B. van, comp. 1978. *Geologic Time Table.* 3d ed. Amsterdam and The Netherlands: Elsevier.

Falla, R. A., R. H. Taylor, and C. Black. 1979. Survey of Dundas Island, Auckland Islands, with particular reference to Hooker's sea lion *(Phocarctos hookeri). N. Z. J. Zool.* 6:347–355.

Fowler, C. W. and J. A. MacMahon. 1982. Selective extinction and speciation: their influences on the structure and functioning of communities and ecosystems. *Am. Nat.* 119:480–498.

Harris, M. P. and T. R. Birkhead. 1985. Breeding ecology of the Atlantic Alcidae. In D. N. Nettleship and T. R. Birkhead, eds., *The Atlantic Alcidae,* pp. 155–204. London: Academic Press.

Harrison, P. 1983. *Seabirds: An Identification Guide*. Boston: Houghton Mifflin Company.

Hendey, Q. B. and C. A. Repenning. 1972. A Pliocene phocid from South Africa. *Ann. S. Afr. Mus.* 59:71–98.

Howard, A. D. 1951. Development of the landscape of the San Francisco Bay Counties. In O. P. Jenkins, ed., *Geologic Guidebook of the San Francisco Bay Counties*, pp. 95–106. San Francisco, Calif.: State Division of Mines, Bulletin 154.

Howard, H. 1958. Miocene sulids of southern California. *Contrib. Sci. Nat. Hist. Mus. Los Angeles Co.* 25:1–15.

Howard, H. 1969. A new avian fossil from Kern County, California. *Condor* 71:68–69.

Howard, H. 1970. A review of the extinct genus, *Mancalla*. *Contrib. Sci. Nat. Hist. Mus. Los Angeles Co.* 203:1–12.

Howard, H. 1982. Fossil birds from the Tertiary marine beds at Oceanside, San Diego County, California, with descriptions of two new species of the genera *Uria* and *Cepphus* (Aves: Alcidae). *Contrib. Sci. Nat. Hist. Mus. Los Angeles Co.* 341:1–15.

Howard, H. 1983. A list of the extinct fossil birds of California. *Bull. So. Calif. Acad. Sci.* 82:1–11.

Howard, H. 1984. Additional records from the Miocene of Kern County, California with the description of a new species of fulmar (Aves: Procellariidae). *Bull. So. Calif. Acad. Sci.* 83:84–89.

Hunter, I., J. P. Croxall, and P. A. Prince. 1982. The distribution and abundance of burrowing seabirds (Procellariiformes) at Bird Island, South Georgia, 1: Introduction and methods. *Br. Antarct. Surv. Bull.* 56:49–67.

Janzen, D. H. and P. S. Martin. 1982. Neotropical anachronisms: the fruits the gomphotheres ate. *Science* 215:19–27.

Jenkins, R. J. F. 1974. A new giant penguin from the Eocene of Australia. *Palaeontology* 17:291–310.

Johnsgard, P. A. 1978. *Ducks, Geese, and Swans of the World*. Lincoln: University of Nebraska Press.

Keller, G. and J. A. Barron. 1981. Integrated planktic foraminiferal and diatom biochronology for the northeast Pacific and the Monterey Formation. In R. E. Garrison and R. G. Douglas, eds., *The Monterey Formation and Related Siliceous Rocks of California*, pp. 43–54. Society of Economic Paleontologists and Mineralogists.

King, J. E. 1983a. *Seals of the World*. British Museum (Natural History) London.

King, J. E. 1983b. The Ohope Skull—a new species of Pleistocene sealion from New Zealand. *N. Z. J. Mar. Fresh. Res.* 17:105–120.

Kohl, R. F. 1974. A new Late Pleistocene fauna from Humboldt County, California. *The Veliger* 17:211–219.

LeBoeuf, B. J. and M. L. Bonnell. 1980. Pinnipeds of the California Islands: abundance and distribution. In D. M. Power, ed., *The California Islands: Proceedings of a Multidisciplinary Symposium*, pp. 475–493. Santa Barbara, Calif.: Santa Barbara Museum of Natural History.

Lindberg, D. R. and M. G. Kellogg. 1982. Bathymetric anomalies in the Neogene fossil record: the role of diving marine birds. *Paleobiology* 8:412–407.

Maurer, B. A. 1984. Interference and exploitation in bird communities. *Wilson Bull.* 96:380–395.

McCann, T. S. 1980. Territoriality and breeding behavior of adult male Antarctic fur seal, *Arctocephalus gazella*. *J. Zool. Lond.* 192:295–310.

Miller, L. 1925. Avian remains from the Miocene of Lompoc, California. *Carnegie Institute of Washington* 349:107–117.

Miller, L. 1961. Birds from the Miocene of Sharktooth Hill, California. *Condor* 63:399–402.

Mitchell, E. D., Jr. 1965. Morphology of a Miocene sea lion. *Geol. Soc. Amer. Spec. Pap.* 87:218.

Mitchell, E. D., Jr. 1966. The Miocene pinniped *Allodesmus. Univ. Calif. Publ. Geol. Sci.* 61:1–105.

Mitchell, E. D., Jr. and R. H. Tedford. 1973. The Enaliarctinae. A new group of extinct aquatic Carnivora and a consideration of the origin of the Otariidae. *Bull. Amer. Mus. Nat. Hist.* 151:201–284.

Morejohn, G. V. 1976. Evidence of the survival to Recent times of the extinct flightless duck *Chendytes lawi* Miller. *Smith. Contrib. Paleo.* 27:207–211.

Muizon, C. de. 1978. *Arctocephalus (Hydrarctos) lomasiensis*, subgen. nov. et nov. sp., un nouvel Otariidae du Mio-Pliocenène de Sacaco (Pérou). *Bulletin de l'Institut Francais d'Etudes andines* 7:169–188.

Muizon, C. de. 1981. Les vertébrés fossiles de la formation Pisco (Pérou), 1: Deux nouveaux Monachinae (Phocidae, Mammalia) du Pliocène de Sud-Sacaco. *Travaux de l'Institut francais d'Etudes andines* 22. (Recherche sur les grandes civilisation mémoire 6, Paris).

Muizon, C. de and M. Bond. 1982. Le Phocidae (Mammalia) miocène de la formation Paraná (Entre Rios, Argentine). *Bulletin du Muséum national d'Histoire naturelle*, Paris, série 4(4), section C:165–207.

Muizon, C. de and Q. B. Hendey. 1980. Late Tertiary seals of the South Atlantic Ocean. *Ann. S. Afr. Mus.* 82:91–128.

Mullineaux, L. S. and M. J. Westberg-Smith. 1986. Radiolarians as paleoceanographic indicators in the Miocene Monterey Formation, upper Newport Bay, California. *Micropaleontology* 32:48–71.

Olson, S. L. 1985a. An early Pliocene marine avifauna from Duinefontein, Cape Province, South Africa. *Ann. S. Afr. Mus.* 95:147–164.

Olson, S. L. 1985b. The fossil record of birds. In D. Farner, J. King, and K. C. Parkes, ed., *Avian Biology*, 8:79–256. New York: Academic Press.

Olson, S. L. 1986. A replacement name for the fossil penguin *Microdytes* Simpson (Aves: Spheniscidae). *J. Paleo.* 60:785.

Olson, S. L. and Y. Hasegawa. 1979. Fossil counterparts of giant penguins from the North Pacific. *Science* 206:688–689.

Olson, S. L. and H. F. James. 1982a. Fossil birds from the Hawaiian Islands: evidence for wholesale extinction by man before western contact. *Science* 217:633–635.

Olson, S. L. and H. F. James. 1982b. Prodromus of the fossil avifauna of the Hawaiian Islands. *Smith. Contrib. Zool.* 365:1–59.

Payne, M. R. 1977. Growth of a fur seal population. *Phil. Trans. R. Soc. Lond. B.* 279:67–79.

Power, D. M. 1980. Introduction. In D. M. Power, ed., *The California Islands: Proceedings of a Multidisciplinary Symposium*, pp. 1–4. Santa Barbara, Calif.: Santa Barbara Museum of Natural History.

Pregill, G. K. and S. L. Olson. 1981. Zoogeography of West Indian vertebrates in relation to Pleistocene climatic cycles. *Ann. Rev. Ecol. Syst.* 12:75–98.

Rand, R. W. 1951. Birds breeding on Seal Island (False Bay, Cape Province). *Ostrich* 22:94–103.

Rand, R. W. 1952. The birds of Hollamsbird Island, South West Africa. *Ibis* 94:452–457.

Rand, R. W. 1954. Notes on the birds of Marion Island. *Ibis* 96:173–206.

Ray, C. E. 1976. Geography of phocid evolution. *Syst. Zool.* 25:391–406.

Repenning, C. A. 1976. Adaptive evolution of Sea Lions and Walruses. *Syst. Zool.* 25:375–390.

Repenning, C. A., C. E. Ray, and D. Grigorescu. 1979. Pinniped biogeography. In J. Gray and A. J. Boucot, eds., *Historical Biogeography, Plate Tectonics, and the Changing Environment*, pp. 357–369. Corvallis: Oregon State University Press.

Repenning, C. A. and R. H. Tedford. 1977. Otarioid seals of the Neogene. *U. S. Geol. Surv. Prof. Paper* 992:i–vi, 1–93.

Ricklefs, R. E. and G. W. Cox. 1972. Taxon cycles in the West Indian avifauna. *Am. Nat.* 106:195–219.

Schoener, T. W. 1983. Field experiments on interspecific competition. *Am. Nat.* 122:240–285.

Shaughnessy, P. D. 1980. Influence of Cape fur seals on jackass penguin numbers on Sinclair Island. *S. Afr. J. Wild. Res.* 10:18–21.

Shaughnessy, P. D. 1984. Historical population levels of seals and seabirds on islands off southern Africa, with special reference to Seal Island, False Bay. *Investl. Rep. Sea Fish. Res. Inst. S. Afr.* 127:1–61.

Simpson G. G. 1965. New record of a fossil penguin in Australia. *Proc. Roy. Soc. Victoria* 79:91–93.

Simpson, G. G. 1970. Ages of fossil penguins in New Zealand. *Science* 168:361–362.

Simpson, G. G. 1971. Review of fossil penguins from Seymour Island. *Proc. R. Soc. Lond. B.* 178:357–387.

Simpson, G. G. 1975. Fossil penguins. In B. Stonehouse, ed., *The Biology of Penguins*, pp. 19–41. London: Macmillan Press.

Simpson, G. G. 1976. *Penguins Past and Present, Here and There*. New Haven, Conn.: Yale University Press.

Simpson, G. G. 1981. Notes on some fossil penguins, including a new genus from Patagonia. *Ameghiniana* 18:266–272.

Sowls, A. L., A. R. DeGange, J. W. Nelson, and G. S. Lester. 1980. *Catalog of California Seabird Colonies*. Washington, D.C.: U.S. Department of the Interior, Fish and Wildlife Service, Biological Services Program. FWS/OBS-37/80.

Sowls, A. L., S. A. Hatch, and C. J. Lensink. 1978. *Catalog of Alaskan Seabird Colonies*. Washington, D.C.: U.S. Department of the Interior, Fish and Wildlife Service. FWS/OBS-78/78.

Steadman, D. W. and S. L. Olson. 1985. Bird remains from an archaeological site on Henderson Island, South Pacific: Man-caused extinctions on an "uninhabited" island. *Proc. Nat. Acad. Sci.* 82:6191–6195.

Stirling, I. 1975. Factors affecting the evolution of social behavior in the pinnipedia. *Rapp. P.-v. Reun. Con. int. Explor. Mer.* 169:205–212.

Stonehouse, B. 1967. The general biology and thermal balances of penguins. *Adv. Ecol. Res.* 4:673–675.

Tonni, Eduardo P. 1980. The present state of knowledge of the Cenozoic birds of Argentina. *Contrib. Sci. Nat. Hist. Mus. Los Angeles Co.* 330:105–114.

Vail, P. R. and J. Hardenbol. 1979. Sea-level changes during the Tertiary. *Oceanus* 22:71–79.

Vedder, J. G. and D. G. Howell. 1980. Topographic evolution of the southern California Borderland during the late Cenozoic time. In D. M. Power, ed., *The California Islands: Proceedings of a Multidisciplinary Symposium*, pp. 7–27. Santa Barbara, Calif.: Santa Barbara Museum of Natural History.

Warheit, K. I., D. R. Lindberg, and R. J. Boekelheide. 1984. Pinniped disturbance lowers reproductive success of Black Oystercatcher *Haematopus bachmani* (Aves). *Mar. Ecol. Prog. Ser.* 17:101–104.

Wetmore, A. 1930. Fossil bird remains from the Temblor Formation near Bakersfield, California. *Proc. Calif. Acad. Sci.* 19 (ser. 4):85–93.

Wilson, G. J. 1979. Hooker's sea lions in southern New Zealand. *N. Z. J. Mar. Fresh. Res.* 13:373–375.

■ Species Index

■ Subject Index

■ Contributors

David W. Au
National Marine Fisheries Service
La Jolla, California, USA 92038

Joanna Burger
Department of Biological Sciences
Rutgers University
Piscataway, New Jersey, USA 08855

Anthony DeGange
U.S. Fish and Wildlife Service
Anchorage, Alaska

Kenneth Ensor
Department of Zoology
University of Glasgow
Glasgow, Scotland G12 BQQ

Robert W. Furness
Department of Zoology
University of Glasgow
Glasgow, Scotland G12 BQQ

Anne V. Hudson
Department of Zoology
University of Glasgow
Glasgow, Scotland G12 BQQ

Kees Hulsman
School of Australian Environmental Studies
Griffith University
Nathan, Queensland, Australia 4111

Linda L. Jones
National Marine Mammal Laboratory
Northwest and Alaska Fisheries Center
Seattle, Washington, USA 98115

David R. Lindberg
Museum of Paleontology
University of California
Berkely, California, USA 94720

Raymond J. Pierotti
Department of Zoology, Birge Hall
University of Wisconsin
Madison, WI 53706

Robert L. Pitman
National Marine Fisheries Service
Southwest Fisheries Center
La Jolla, California, USA 92038

Carl Safina
National Audubon Society
Islip, New York, USA 11751

Kenneth I. Warheit
Department of Paleontology
University of California
Berkely, California, USA 94704